SPRINGER TRACTS IN MODERN PHYSICS

Ergebnisse
der exakten Natur-
wissenschaften

Volume **62**

Editor: G. Höhler

Editorial Board: P. Falk-Vairant S. Flügge J. Hamilton
F. Hund H. Lehmann E. A. Niekisch W. Paul

Springer-Verlag Berlin Heidelberg GmbH 1972

Manuscripts for publication should be adressed to:

G. Höhler, Institut für Theoretische Kernphysik der Universität, 75 Karlsruhe 1, Postfach 6380

Proofs and all correspondence concerning papers in the process of publication should be addressed to:

E. A. Niekisch, Kernforschungsanlage Jülich, Institut für Technische Physik, 517 Jülich, Postfach 365

ISBN 978-3-662-15578-3   ISBN 978-3-540-37165-6 (eBook)

DOI 10.1007/978-3-540-37165-6

© by Springer-Verlag Berlin Heidelberg 1972

Softcover reprint of the hardcover 1st edition 1972

Originally published by Springer-Verlag Berlin Heidelberg New York in 1972

Library of Congress Catalog Card Number 25-9130

Photon-Hadron Interactions I

International Summer Institute
in Theoretical Physics
DESY, July 12—24, 1971

Contents

Canonical Light-Cone Commutators
and Their Applications

Roman Jackiw*

Contents

I. Introduction

I shall report here on recent work done by several colleagues[1] and me where we succeeded in exhibiting the canonical structure of the commutators of currents on the light cone, and applied it to several interesting physical problems. Two reasons may be advanced at the present time for studying this topic.

(1) It has been appreciated already in the early days of current algebra, that the fixed mass sum rules, the most famous of which is due to Dashen, Fubini and Gell-Mann [1, 2], which are usually derived with the help of the unreliable $p \to \infty$ technique [3], are in fact equivalent to appropriate light-cone commutators [4].

---

* A. P. Sloan Fellow.
[1] This work was done in collaboration with J. M. Cornwall, D. Discus and V. Teplitz.

(2) Recently it has been shown that the dramatic MIT-SLAC electro-production experiments [5] measure directly the commutator of electromagnetic currents on the light cone [6].

The following notation will be used. For the coordinate x^μ and for all other vector or tensor quantities, we define the $\pm$ components by

$$x^\pm = 2^{-1/2}(x^0 \pm x^3). \tag{I-1}$$

The remaining components will be denoted by $i, i = 1, 2$, or by the subscript $\perp$. The metric tensor now is $g^{++} = g^{--} = g^{12} = g^{21} = 0$; $g^{+-} = g^{-+} = -g^{11} = -g^{22} = 1$. Consequently $x^2 = 2x^+ x^- - x_\perp^2$.

Consider the commutator of two currents.

$$C_{ab}^{\mu\nu}(x) = [J_a^\mu(x), J_b^\nu(0)]. \tag{I-2}$$

In a given field theory, the usual canonical formalism permits one to evaluate the *equal time commutator*, $C_{ab}^{\mu\nu}(x)|_{x^0=0}$, without solving the theory. I shall show how similarly one can compute the *light cone commutator*, $C_{ab}^{\mu\nu}(x)|_{x^+=0}$, without unraveling the dynamics. The reason for the nomenclature can be given. Since $C_{ab}^{\mu\nu}(x)$ vanishes for negative x^2, and since $x^2 = 2x^+ x^- - x_\perp^2 = -x_\perp^2$ when $x^+ = 0$, knowledge of $C_{ab}^{\mu\nu}(x)$ at the point $x^+ = 0$ is equivalent to the knowledge of $C_{ab}^{\mu\nu}(x)$ on the light cone, $x^2 = 0$. The relation of the present results to other investigations is as follows. Fritsch and Gell-Mann [7] independently and simultaneously suggested that the *most singular* contribution to $C_{ab}^{\mu\nu}(x)$ near $x^2 = 0$ might be as in the free-field theory, where of course it is completely calculable. Subsequently Gross and Treiman [8] verified the Fritsch-Gell-Mann [7] conjecture in various interacting theories. For purposes of studying the deep-inelastic, "scaling" region [9], both our approach and the Fritsch-Gell-Mann [7] technique give identical results. However, as will be seen below, the determination of the fixed mass sum rules requires the *full* commutator at $x^+ = 0$, not just the most singular part at $x^2 = 0$.

II. Light Cone Current Commutators

Since, as we have argued, light cone current commutators are of interest both for experimental and theoretical physics, we now show how one might calculate them [10]. Consider first a conserved, internal symmetry current J_a^μ. The time independent charge is given by $Q_a = \int d^3x J_a^0(x)$, and the assumption that the current transforms in a known fashion in group space implies

$$[Q_a, J_a^\mu(0)] = i f_{abc} J_c^\mu(0). \tag{II-1}$$

It is not hard to show that for *conserved* currents an alternate formula for Q_a may be given.

$$Q_a = \int d^2 x_\perp d x^- J_a^+(x). \tag{II-2}$$

Of course Q_a is independent of the unintegrated variable x^+. From (II-1) and (II-2) it now follows that

$$[J_a^+(x), J_b^\mu(0)]_{x^+=0} = i f_{abc} J_c^\mu(0) \, \delta^2(x_\perp) \, \delta(x^-)$$
$$+ \partial_- S_{ab}^\mu(x) + \partial_i S_{ab}^{i\mu}(x). \tag{II-3}$$

The terms in (II-3) which disappear upon integration over x^- and $x_\perp$ are not unlike the Schwinger terms that are present in selected equal time commutators. Indeed by taking vacuum expectation values, one can show that they *must* be present, though of course the vacuum matrix element is not sensitive to their c or q number character. However in one important respect these structures are different from what has been encountered previously. $S_{ab}^\mu(x)$ and $S_{ab}^{i\mu}(x)$ are not, in general, local in x^-, though they do possess this properly in $x_\perp$. The reason for this is the following. The commutator must vanish for $x^2 = 2x^+ x^- - x_\perp^2 < 0$. When $x^+ = 0$, as it is in (II-3), $x^2 = -x_\perp^2$, and by causality, the commutator must be local on $x_\perp$ (i.e. it has support only at $x_\perp = 0$). However no constraint on x^- is imposed. Therefore we expect the objects on the right hand side of (II-3) to have an arbitrary dependence on x^-. Such quantities are called *bilocal* operators.

The bilocal operators are model dependent, and at the present time there is no *a priori* way of calculating them. However, as we shall demonstrate repeatedly, they are measurable, physical quantities, and their form must be specified, if we wish to understand various and diverse physical phenomena. As will be demonstrated below, they are known to have non-zero connected matrix elements; hence they are q numbers. It will be shown presently that the $p \to \infty$ technique of current algebra would predict their vanishing; hence we have the possibility of rectifying that unreliable method, if we can specify the bilocal operators. It is this task which we now embark upon.

A. Light-Cone Quantization

The reason it was possible to give a plausible model for *equal time* commutators without solving any of the dynamical equations of quantum field theory, was of course due to the fact that equal time commutators for fields form a boundary condition on the theory. They are assumed to be of a definite canonical form which can be given without solving the theory. Consequently for any function of the fields, like a current, the appropriate commutators can be computed. It turns out that something similar may be done for light cone commutators. Rather than quantizing the theory on a space like surface – which corresponds to the usual equal time quantization – one may equivalently quantize on a light-like surface. With the latter technique, light cone commutators emerge canonically [11].

In order to obtain familiarity with these ideas, let us begin by examining a simple scalar theory given by the Lagrangian

$$\mathscr{L} = \tfrac{1}{2}\, \partial_\mu \varphi \, \partial^\mu \varphi - \tfrac{1}{2}\, \mu^2\, \varphi^2 + \tfrac{1}{4}\, \lambda \varphi^4. \tag{II-4}$$

In the usual way, this leads to the equation of motion

$$\Box \varphi + \mu^2\, \varphi = \lambda \varphi^3. \tag{II-5}$$

However the quantum theory is not specified only by (II-5). One also postulates the following equal time commutation relations

$$i\,[\varphi(x),\, \varphi(0)]_{x^0=0} = 0, \tag{II-6a}$$

$$i\,[\partial_0\, \varphi(x),\, \varphi(0)]_{x^0=0} = \delta^3(\boldsymbol{x}), \tag{II-6b}$$

$$i\,[\partial_0\, \varphi(x),\, \partial_0\, \varphi(0)]_{x^0=0} = 0 \tag{II-6c}$$

where the special role of $\partial_0\, \varphi$ follows from the fact that it is the canonical momentum $\partial_0\, \varphi = \partial \mathscr{L}/\delta \partial_0\, \varphi$. This scheme is the quantum analogue of the classical (non-quantized) procedure of solving the partial differential (II-5) with the specification of the Cauchy initial value data: $\varphi(x)$ and $\partial_0\, \varphi(x)$ on the space like surface $x^0 = 0$.

Let us now rewrite things in terms of the light cone variables. Of course the Lagrange function and the equation of motion are equivalent.

$$\mathscr{L} = \partial_-\, \varphi\, \partial_+\, \varphi + \tfrac{1}{2}\, \partial_i\, \varphi\, \partial^i\, \varphi - \tfrac{1}{2}\, \mu^2 \varphi^2 + \tfrac{1}{4}\, \lambda \varphi^4, \tag{II-7}$$

$$2\, \partial_+\, \partial_-\, \varphi + \partial_i\, \partial^i\, \varphi + \mu^2\, \varphi = \lambda \varphi^3. \tag{II-8}$$

The new element comes in if we define the canonical momentum to be $\delta \mathscr{L}/\delta \partial_+\, \varphi = \partial_-\, \varphi$. Consequently we postulate that

$$i\,[\partial_-\, \varphi(x),\, \varphi(0)]_{x^+=0} = \tfrac{1}{2}\, \delta(x^-)\, \delta^2(\boldsymbol{x}_\perp). \tag{II-9a}$$

This may be integrated and gives the basic light cone commutator

$$i\,[\varphi(x),\, \varphi(0)]_{x^+=0} = \tfrac{1}{4}\, \varepsilon(x^-)\, \delta^2(\boldsymbol{x}_\perp) \tag{II-9b}$$

where $\varepsilon(x) = -\varepsilon(-x) = 1$ for $x > 0$. This procedure is the quantum analog of solving the classical theory by specifying the initial data on the light like surface $x^+ = 0$. It can be shown that classically a unique solution is obtained by specifying only φ. The knowledge of derivatives of φ is unnecessary [12].

Therefore (II-9b) is the fundamental light cone commutator in this theory, and light cone commutators of all other functions of $\varphi(x)$ may be now computed without solving the theory. There are peculiar features of this quantization scheme that should be commented upon. Note that the commutator between field and momentum (II-9a) has an unexpected factor of $\frac{1}{2}$. Furthermore according to (II-9b) fields do not commute with themselves, and it is also clear that canonical momenta are non-commuting. This difference from the state of affairs in equal time quantization arises from the fact that the canonical momentum $\pi(x)$ is related by an equation of *constraint* to the canonical field: $\pi(x) = \partial_-\varphi(x)$. In the present scheme equations of motion involve $+$ derivatives; the $+$ coordinate plays the role of time. Consequently the canonical formalism is more complicated than at equal times, and careful investigation shows that the *Ansatz* (II-9) is correct. A related phenomenon is that the light cone method has changed a second order differential equation in time, to a first order equation in $+$. Therefore the present scheme is somewhat analogous to the conventional quantization of Fermion theories, which are also first order.

The question now arises whether or not the light cone theory is the same as the equal time theory, or whether it is different; i.e. is the S matrix the same for both schemes. This is an important question, for if the two theories are different, then we cannot assume the *simultaneous* validity of the conventional equal time commutators and the new light cone commutators. To answer this question with complete certainty would require solving the two theories, and comparing the results – a task impossible to carry out at present. Hence we must content ourselves with a partial answer which relies on various formal properties of the two theories. The following four facts are offered as evidence that the two methods of quantization result in the same physical theory.

1. It can be shown that the free theories are the same [11].

2. If the light cone theory is developed to the point of computing S matrix elements, then one encounters instead of the usual Feynman-Dyson rules, the Weinberg "$p \to \infty$" [13] rules, which are supposed to be equivalent [11].

3. In the light cone theory, Green's functions are ordered along the $+$ direction, rather than along the time direction. However causality requires the two to be the same, up to seagulls. To see this, consider

$$T_0(x) = [A(x), B(0)]\,\theta(x^0), \qquad\qquad \text{(II-10a)}$$

$$T_+(x) = [A(x), B(0)]\,\theta(x^+). \qquad\qquad \text{(II-10b)}$$

The only place that (II-10a) and (II-10b) appear to be unequal is when $x^0 > 0$ and $x^+ < 0$, or $x^0 < 0$ and $x^+ > 0$. However in this region x^2 is spacelike, and the commutator function vanishes by causality [14].

4. The *vacuum expectation value* of light cone commutators can be calculated in the usual framework. The results are consistent with the canonically postulated forms. For example, consider

$$i \langle 0| [\varphi(x), \varphi(0)] |0\rangle = i \int_0^\infty d a^2 \varrho(a^2) \Delta(x|a^2) \qquad \text{(II-11)}$$

where $\varrho(a^2)$ is some spectral function, and $\Delta(x|a^2)$ is the free-field commutator function.

$$\Delta(x|a^2) = (2\pi)^{-3} \int d^4 k \, \varepsilon(k^0) \, \delta(k^2 - a^2) \, e^{-ikx}. \qquad \text{(II-12)}$$

In the usual way, we have from the equal time commutators the result that

$$\int_0^\infty d a^2 \, \varrho(a^2) = 1. \qquad \text{(II-13)}$$

To calculate the light cone commutator, we observe that from (II-12) it follows that

$$\Delta(x|a^2)|_{x^+ = 0} = (-i/4) \, \varepsilon(x^-) \, \delta^2(\boldsymbol{x}_\perp). \qquad \text{(II-14)}$$

Consequently, we have with the help of (II-13)

$$i \langle 0| [\varphi(x), \varphi(0)] |0\rangle|_{x^+ = 0} = \tfrac{1}{4} \varepsilon(x^-) \, \delta^2(\boldsymbol{x}_\perp) \qquad \text{(II-15)}$$

which reproduces the canonically postulated (II-9b), as far as vacuum expectation values are concerned. Other evaluations of vacuum expectation values always lead to the same result – the light cone method is consistent with and equivalent to the equal time method.

Supported by the above four partial arguments, as well as by the fact that no conflict has been found between the two methods, we shall assume that the form of light cone commutators evaluated in the light cone quantization scheme is valid simultaneously with the equal time commutators.

To give a form for the $SU(3) \times SU(3)$ current commutators, we turn to a model which has previously served as an inspiration for equal time commutators: the quark model. Since we wish to present results which are possibly true even in an interacting field theory, we shall assume that the quarks interact with a vector gluon – the end result does not depend on the nature of the gluon-quark interaction, as long as it is a non-derivative one. Thus we are led to the Lagrangian

$$\mathscr{L} = \bar{\psi} [((i/2) \, \overset{\leftrightarrow}{\partial}_\mu - g B_\mu) \gamma^\mu - M] \psi$$
$$- \tfrac{1}{4} F^{\mu\nu} F_{\mu\nu} \qquad \text{(II-16)}$$
$$F^{\mu\nu} = \partial^\mu B^\nu - \partial^\nu B^\mu.$$

We have set the gluon (B^μ) mass to zero in order that the computations be simple — the end results with which we deal are insensitive to this simplification. The quark field ψ has mass M. If we wish to introduce $SU(3)$ symmetry breaking we would allow M to be a mass matrix, though for the present, we do not do this. The light cone quantization of this model may now be taken over from the published literature [11]. Various peculiar features of this scheme are the following. The 4 component Fermion field ψ cannot be viewed as a canonically independent field. Rather a projection of it $\psi_+ \equiv P_+ \psi, P_+ = \frac{1}{2}\gamma^- \gamma^+$ is independent; while the other projection $\psi_- \equiv P_- \psi, P_- = \frac{1}{2}\gamma^+ \gamma^-$ is a dependent field. The canonical dependence is expressed by the following equation of constraint, which may be shown to hold in this theory.

$$(i\,\partial_- - g B^+)\,\psi_- = \tfrac{1}{2}\left[(i\,\partial_j - g B_j)\,\gamma^j + M\right]\gamma^+ \psi_+ . \qquad \text{(II-17a)}$$

Since B^μ is a *massless* vector meson field, there exists a gauge freedom, which may be exploited to set B^+ to zero [15]. Then (II-17a) is easily integrated, to give a formula, at fixed x^+, for ψ_- in terms of the canonically independent fields B_j and ψ_+.

$$\psi_-(x) = (-i/4)\int \mathrm{d}\xi\, \varepsilon(x^- - \xi)$$
$$\cdot \{[i\,\partial_j - g B_j(x^+, \xi, \boldsymbol{x}_\perp)]\,\gamma^j + M\}\,\gamma^+ \psi_+(x^+, \xi, \boldsymbol{x}_\perp) . \qquad \text{(II-17b)}$$

The canonical commutators are

$$[B^i(x), B^j(0)]_{x^+ = 0} = (i/4)\,g^{ij}\varepsilon(x^-)\,\delta^2(\boldsymbol{x}_\perp) , \qquad \text{(II-18)}$$

$$\{\psi_+(x), \psi_+^*(0)\}_{x^+ = 0} = 2^{-1/2}\,P_+\,\delta(x^-)\,\delta^2(\boldsymbol{x}_\perp) . \qquad \text{(II-19)}$$

The operators ψ_+ and ψ_+^* anticommute with themselves, and commute with the B^i. Another commutator which we shall need is the one between ψ_+^* and ψ_-. We have from (II-17) and (II-19)

$$\{\psi_-(x), \psi_+^*(x')\}_{x^+ = x'^+}$$
$$= -(i/4\sqrt{2})\,\varepsilon(x^- - x'^-)\,\{[i\,\partial_j - g B_j(x')]\,\gamma^j + M\}\,\gamma^+ \delta^2(\boldsymbol{x}_\perp - \boldsymbol{x}'_\perp) . \qquad \text{(II-20)}$$

It is also possible to give other commutators, e.g. between ψ_- and ψ_-^*. These are extremely complicated, and fortunately we have no need of them. Note that the commutator (II-20) makes reference to the interaction.

B. Derivation of Current Commutators

The current commutators may now be calculated. For the present, considerations of internal symmetry are ignored, and the current is defined by

$$J^\mu = \bar{\psi}\gamma^\mu \psi . \qquad \text{(II-21a)}$$

In terms of the fields ψ_+ and ψ_- the formula for the current is as follows

$$J^+ = 2^{1/2}\,\psi_+^*\,\psi_+ \qquad\qquad\qquad\qquad\text{(II-21 b)}$$

$$J^- = 2^{1/2}\,\psi_-^*\,\psi_- \qquad\qquad\qquad\qquad\text{(II-21 c)}$$

$$J^i = 2^{-1/2}\,\psi_-^*\,\gamma^+\gamma^i\psi_+ + 2^{-1/2}\,\psi_+^*\,\gamma^-\gamma^i\psi_- . \qquad\text{(II-21 d)}$$

The commutator of J^+ with itself is quite simple. Only the ψ_+ field is involved whose commutator, (II-19), contains no interaction terms. We find

$$[J^+(x), J^+(y)]_{x^+=y^+} = 0 . \qquad\qquad\qquad\text{(II-22)}$$

For the $[J^+, J^-]$ commutator, we need the more complicated formula (II-20). A rather lengthy computation yields

$$[J^+(x), J^-(y)]_{x^+=y^+}$$
$$= (-i/2)\varepsilon(x^- - y^-)\overline{\psi}(x)\{\gamma^j[i\,\overleftrightarrow{\partial}_j + g\,B_j] + M\}\,P_-\psi(y)\delta^2(\boldsymbol{x}_\perp - \boldsymbol{y}_\perp) \quad\text{(II-23 a)}$$
$$- \tfrac{1}{2}\varepsilon(x^- - y^-)\partial_i^x\{\overline{\psi}(x)\gamma^i P_-\psi(y)\delta^2(\boldsymbol{x}_\perp - \boldsymbol{y}_\perp)\} - \text{h.c.}$$

(The abbreviation h.c. stands for Hermitian conjugate which must be subtracted from the right hand side of (II-23 a).) The commutator is not expressible in terms of the current itself. Like the commutator of the fields it appears to depend explicitly on the coupling to $g\,B^\mu$. However this dependence may be eliminated. By use of the equation (II-17 a), (II-23 a) may be reexpressed by

$$[J^+(x), J^-(y)]_{x^+=y^+}$$
$$= \partial_-^x\{-\tfrac{1}{2}\varepsilon(x^- - y^-)\,\delta^2(\boldsymbol{x}_\perp - \boldsymbol{y}_\perp)\,\overline{\psi}(x)\,\gamma^-\psi(y)\}$$
$$+ \partial_i^x\{-\tfrac{1}{4}\varepsilon(x^- - y)\,\delta^2(\boldsymbol{x}_\perp - \boldsymbol{y}_\perp)\,[\overline{\psi}(x)\,\gamma^i\psi(y) \qquad\text{(II-23 b)}$$
$$- \varepsilon^{ij}\overline{\psi}(x)\,\gamma_j\gamma_5\,\psi(y)]\} - \text{h.c.}$$

We have used the relations $\gamma^i P_- = \tfrac{1}{2}\gamma^i - \tfrac{1}{2}\varepsilon^{ij}\gamma_j\gamma_5$,

$$\gamma_5 \equiv \gamma^0\gamma^1\gamma^2\gamma^3, \qquad \varepsilon^{ij} = -\varepsilon^{ji} = 1 \quad\text{for}\quad i=1, j=2 .$$

The remarkable feature of this final formula is its elegant simplicity. The light cone commutator of these two currents is expressible in terms of bilocal operators, which are straight-forward generalizations of the local currents. These bilocal operators are $\overline{\psi}(x)\,\gamma^\mu\psi(y)$ and $\overline{\psi}(x)\gamma^\mu\gamma_5\psi(y)$. It is clear[2] that the operators enter with $(x-y)^2 = 0$, since $x^+ = y^+$ and

---

[2] Our formulas for the bilocal currents are gauge dependent, since they were calculated in the special gauge $B^+ = 0$. An explicitly gauge independent form is

$$\overline{\psi}(0, x^-, \boldsymbol{0})\,\gamma^\mu\psi(0)\,\exp i g \int_0^{x^-} dy^-\,B^+(0, y^-, \boldsymbol{0});$$

see Gross and Treiman [8].

$x_\perp = y_\perp$. Consequently the non-locality is confined to the $-$ direction. The bilocal terms do not contribute when an integration over x^- and $x_\perp$ is performed. Thus the structure (II-23 b) is of the type (II-3), which we arrived at by general considerations.

Finally a completely similar argument gives

$$
\begin{aligned}
[J^+(x), J^i(y)]_{x^+ = y^+} \\
= \partial^x_- \{ -\tfrac{1}{4}\varepsilon(x^- - y^-)\delta^2(x_\perp - y_\perp)[\bar\psi(x)\gamma^i\psi(y) + \varepsilon^{ij}\bar\psi(x)\gamma_j\gamma_5\psi(y)]\} \\
+ \partial^x_j \{ \tfrac{1}{4}\varepsilon(x^- - y^-)\,\delta^2(x_\perp - y_\perp)\,[g^{ij}\bar\psi(x)\gamma^+\psi(y) \\
+ \varepsilon^{ij}\bar\psi(x)\gamma^+\gamma_5\psi(y)]\} - \text{h.c.}
\end{aligned}
\tag{II-24}
$$

Again the light cone commutator is expressible in terms of bilocal generalizations of the vector and axial vector currents.

Commutators of current components, not involving at least one $+$ component, are considerably more complicated as they require the commutators between the ψ_- and ψ_-^* fields. We shall not concern ourselves with these here.

There exists an alternate method for computing commutators, due to Schwinger [16]. Schwinger's original arguments apply to ordinary quantization, but it is not hard to generalize them to light cone quantization. This "action principle" procedure gives the following formula for the *equal time* commutator of the *time* component of a current, with the current itself.

$$
[J^0(x), J^\mu(y)]_{x^0 = y^0} = -i\,\partial_\alpha\,\delta J^\alpha(x)/\delta A_\mu(y).
\tag{II-25}
$$

Here it is assumed that the current is conserved, and that it couples to an external field A^μ. At the end of the computation A^μ is set to zero. The variation in (II-25) is performed with fixed canonical variables, i.e. J^α is first expressed in terms of canonical coordinates and momenta and the variation is performed only with respect to the residual dependence on A^μ. The analogous formula arising from light cone quantization is

$$
[J^+(x), J^\mu(y)]_{x^+ = y^+} = -i\,\partial_\alpha\,\delta J^\alpha(x)/\delta A_\mu(y).
\tag{II-26}
$$

In many models, J^+ has no dependence on external fields. (In our theory for example $J^+ = \sqrt{2}\,\psi_+^*\psi_+$ and ψ_+ is a canonical variable which is held fixed in the explicit variation.) In that case (II-26) becomes

$$
[J^+(x), J^\mu(y)]_{x^+ = y^+} = -i\,\partial_-\,\delta J^-(x)/\delta A_\mu(y) - i\,\partial_i\,\delta J^i(x)/\delta A_\mu(y).
\tag{II-27}
$$

This formula is exactly of the general form (II-3). An abstract representation for the bilocal operators is given by (II-27): $-i\,\delta J^-(x)/\delta A_\mu(y)$ and $-i\,\delta J^i(x)/\delta A_\mu(x)$, and it is seen that when an integration over x^- and $x_\perp$ is performed, the bilocal terms cease to contribute. Moreover we again recognize that these structures are light-cone generalizations of the ordinary Schwinger term. That object, it will be recalled, is given by $\delta J^\nu(x)/\delta A_\mu(x)$, where the variation is performed at a common value of time [16].

A specific form for the bilocal operators is obtainable only by specifying a model, i.e. by specifying the dependence on A^μ. It is not hard to show that in the quark-gluon model this dependence is such that (II-22), (II-23) and (II-24) are true, regardless of the nature of the gluoan, provided it is not coupled derivatively.

C. Non-Canonical Contributions

Let us take the vacuum expectation value of the current commutator function

$$\langle 0| [J^\mu(x), J^\nu(0)] |0\rangle = (g^{\mu\nu}\square - \partial^\mu \partial^\nu) \int_0^\infty d\lambda^2 \sigma(\lambda^2) \Delta(x|\lambda^2). \quad \text{(II-28)}$$

The representation is conserved, as a consequence of current conservation, and a non-negative spectral function $\sigma(\lambda^2)$ has been introduced. Consider now the $++$ components of (II-28) when $x^+ = 0$. We have from (II-14), a nonvanishing result.

$$\langle 0| [J^+(x), J^+(0)] |0\rangle_{x^+=0}$$

$$= - \partial_- \partial_- \int_0^\infty d\lambda^2 \sigma(\lambda^2) \Delta(x|\lambda^2)|_{x^+=0} \quad \text{(II-29)}$$

$$= \partial_- \partial_- [(i/4) \varepsilon(x^-) \delta^2(x_\perp)] \int_0^\infty d\lambda^2 \sigma(\lambda^2) \neq 0.$$

On the other hand the canonical light cone commutator (II-22) would lead us to expect zero for the right hand side of (II-29). Of course, this is just the ancient problem that naive canonical commutators do not yield the conventional Schwinger term in Fermion theories. Recall that the Schwinger term is given by

$$\langle 0| [J^0(x), J^i(0)] |0\rangle_{x^0=0} = i \partial^i \delta^3(x) \int_0^\infty d\lambda^2 \sigma(\lambda^2) \quad \text{(II-30)}$$

which involves the same spectral integral as (II-29). Hence we learn that the light cone commutator (II-22) must be modified by a non-canonical contribution, just as the equal time commutator. We shall make the very important assumption that all these non-canonical additions are c numbers, and therefore given by their vacuum expectation value. This assumption is *not* true in perturbation theory, as will be seen in Section III below. Thus we set

$$[J^+(x), J^+(y)]_{x^+ = y^+} = (-i/4)\, \partial^x_- \partial^y_- [\varepsilon(x^- - y^-)\, \delta^2(\mathbf{x}_\perp - \mathbf{y}_\perp)\, S]$$

$$S = \int_0^\infty d\lambda^2 \sigma(\lambda^2). \qquad (\text{II-31})$$

It is instructive to examine this commutator in theories where the usual Schwinger term emerges canonically e.g. in scalar electro-dynamics. A straight forward calculation yields (II-31), except S becomes a bilocal operator $S(x|y)$, with $S(x|x)$ coinciding with the usual Schwinger term [17].

Similar modifications are present in other commutators. For the $+ -$ components (II-28) gives

$$\langle 0| [J^+(x), J^-(0)] |0\rangle_{x^+ = 0} \qquad (\text{II-32})$$

$$= (i/8) \int_0^\infty d\lambda^2 (\lambda^2 - \partial_i \partial^i)\, \sigma(\lambda^2)\, \varepsilon(x^-)\, \delta^2(\mathbf{x}_\perp).$$

Apart from the term not involving derivatives, which can be reproduced by the canonical commutators, we find in (II-32) a contribution proportional to $S\, \partial_i \partial^i \delta^2(\mathbf{x}_\perp)$. This second derivative structure has no analogue in the canonical result, hence the latter must be modified.

$$[J^+(x), J^-(0)]|_{x^+ = y^+}$$
$$= \partial^x_- \{ -\tfrac{1}{2}\varepsilon(x^- - y^-)\, \delta^2(\mathbf{x}_\perp - \mathbf{y}_\perp)\, \overline{\psi}(x)\, \gamma^-\psi(y) \}$$
$$+ \partial^x_i \{ -\tfrac{1}{4}\varepsilon(x^- - y^-)\, \delta^2(\mathbf{x}_\perp - \mathbf{y}_\perp)[\overline{\psi}(x)\gamma^i\psi(y) - \varepsilon^{ij}\overline{\psi}(x)\gamma_j\gamma_5\psi(y)] \} \qquad (\text{II-33})$$
$$- \text{h.c.} - (i/8)\, \partial^x_i \partial^i_x \{\varepsilon(x^- - y^-)\, \delta^2(\mathbf{x}_\perp - \mathbf{y}_\perp)\, S\}.$$

For the $+i$ component (II-28) gives

$$\langle 0| [J^+(x), J^i(0)] |0\rangle_{x^+ = 0} = (i/4)\, \partial_- \partial^i \varepsilon(x^-)\, \delta^2(\mathbf{x}_\perp)\, S \neq 0 \quad (\text{II-34})$$

while the canonical form gives zero. Hence this commutator is also modified

$$[J^+(x), J^i(y)]_{x^+ = y^+}$$
$$= \partial^x_- \{ -\tfrac{1}{4}\varepsilon(x^- - y^-)\delta^2(\mathbf{x}_\perp - \mathbf{y}_\perp)[\overline{\psi}(x)\gamma^i\psi(y) + \varepsilon^{ij}\overline{\psi}(x)\gamma_j\gamma_5\psi(y)] \}$$
$$+ \partial^x_j \{\tfrac{1}{4}\varepsilon(x^- - y^-)\delta^2(\mathbf{x}_\perp - \mathbf{y}_\perp)\, [g^{ij}\overline{\psi}(x)\, \gamma^+\psi(y) \qquad (\text{II-35})$$
$$+ \varepsilon^{ij}\overline{\psi}(x)\gamma^+\gamma_5\psi(y)] \} - \text{h.c.} + (i/4)\, \partial^x_- \partial^i_x \{\varepsilon(x^- - y^-)\delta^2(\mathbf{x}_\perp - \mathbf{y}_\perp)S\}.$$

In this discussion we have pretended that S is finite. If it is quadratically divergent, which is the scale invariant result, then in addition to the exhibited, quadratically divergent vacuum singularities, there are also well defined, non-infinite terms involving higher derivatives of $\delta^2(\mathbf{x}_\perp)$. We shall not concern ourselves with these objects.

D. $SU(3)$ Generalization

The $SU(3)$ generalization of the previous results is obtained easily. The current is given by

$$V_a{}^\mu(x) = \bar\psi(x)\,\gamma^\mu \tfrac{1}{2}\lambda_a \psi(x)\,. \tag{II-36}$$

We shall also need the axial current

$$A_a^\mu(x) = i\,\bar\psi(x)\,\gamma^\mu \gamma_5 \tfrac{1}{2}\lambda_a \psi(x)\,. \tag{II-37}$$

The internal symmetry matrices satisfy

$$\lambda_a \lambda_b = (i f_{abc} + d_{abc})\,\lambda_c\,. \tag{II-38}$$

In deriving the current commutators, we allow for non-conservation of the currents by introducing the mass matrix M in the Lagrangian (II-16). The commutators are the following

$$\begin{aligned}
[V_a{}^+(x), V_b{}^+(y)]_{x^+ = y^+} &= i f_{abc} V_c{}^+(x)\,\delta(x^- - y^-)\,\delta^2(\mathbf{x}_\perp - \mathbf{y}_\perp)\\
&\quad - (i/4)\,\delta_{ab}\,\partial_-^x\,\partial_-^y\,[\varepsilon(x^- - y^-)\,\delta^2(\mathbf{x}_\perp - \mathbf{y}_\perp)\,S]\,,
\end{aligned} \tag{II-39}$$

$$\begin{aligned}
&[V_a{}^+(x), V_b{}^-(y)]_{x^+ = y^+} - i f_{abc} V_c{}^-(x)\,\delta(x^- - y^-)\,\delta^2(\mathbf{x}_\perp - \mathbf{y}_\perp)\\
&= -\tfrac{1}{4}\{i f_{abc} + d_{abc}\}\,\{\partial_-^x\,[\varepsilon(x^- - y^-)\,\delta^2(\mathbf{x}_\perp - \mathbf{y}_\perp)\,V_c{}^-(x|y)]\\
&\quad + \tfrac{1}{2}\partial_i^x[\varepsilon(x^- - y^-)\,\delta^2(\mathbf{x}_\perp - \mathbf{y}_\perp)\,(V_c^i(x|y) + i\varepsilon^{ij} A_{jc}(x|y)]\}\\
&\quad + \tfrac{1}{16} i\varepsilon(x^- - y^-)\,\delta^2(\mathbf{x}_\perp - \mathbf{y}_\perp)\,\bar\psi(x)\,\gamma^+ \gamma^-\,\Lambda_{ab}\psi(y) - \text{h.c.}\\
&\quad - \tfrac{1}{8} i\delta_{ab}\,\partial_i^x\,\partial_x^i\,\{\varepsilon(x^- - y^-)\,\delta^2(\mathbf{x}_\perp - \mathbf{y}_\perp)\,S\}\,,
\end{aligned} \tag{II-40}$$

$$\begin{aligned}
&[V_a{}^+(x), V_b^i(y)]_{x^+ = y^+} - i f_{abc} V_c^i(x)\,\delta(x^- - y^-)\,\delta^2(\mathbf{x}_\perp - \mathbf{y}_\perp)\\
&= -\tfrac{1}{8}\{i f_{abc} + d_{abc}\}\,\{\partial_-^x\,[\varepsilon(x^- - y^-)\delta^2(\mathbf{x}_\perp - \mathbf{y}_\perp)\,(V_c^i(x|y) - i\varepsilon^{ij} A_{jc}(x|y))]\\
&\quad + \partial_j^x[\varepsilon(x^- - y^-)\,\delta^2(\mathbf{x}_\perp - \mathbf{y}_\perp)\,(-g^{ij} V_c{}^+(x|y) + i\varepsilon^{ij} A_c{}^+(x|y))]\}\\
&\quad + \tfrac{1}{16} i\varepsilon(x^- - y^-)\,\delta^2(\mathbf{x}_\perp - \mathbf{y}_\perp)\,\bar\psi(x)\,\gamma^+ \gamma^i \Lambda_{ab}\psi(y) - \text{h.c.}\\
&\quad + \tfrac{1}{4} i\delta_{ab}\,\partial_-^x\,\partial_x^i\,\{\varepsilon(x^- - y^-)\,\delta^2(\mathbf{x}_\perp - \mathbf{y}_\perp)\,S\}\,.
\end{aligned} \tag{II-41}$$

We have introduced the bilocal generalizations of the vector and axial vector current

$$V_a^\mu(x|y) = \bar\psi(x)\,\gamma^\mu \tfrac{1}{2}\lambda_a \psi(y) \tag{II-42}$$

$$A_a^\mu(x|y) = i\bar\psi(x)\,\gamma^\mu \gamma_5 \psi(y)\,. \tag{II-43}$$

These enter the commutation relations with $(x - y)^2 = 0$, $x^- - y^- \neq 0$. The term involving Λ_{ab} is an internal symmetry operator given by $\Lambda_{ab} = [M, \lambda_a]\,\lambda_b$, i.e. it probes the non conservation of the current V_a^μ. It is not hard to see how the commutators with the axial current work out, but we do not present them here. In Section IV the model dependence of these formulas will be discussed.

E. BJL Theorem on the Light Cone

We conclude this Section by presenting a method for computing light cone commutators directly from Green's functions. The method is analogous to the usual BJL [18, 19] technique which allows one to determine an equal time commutator. Consider the x^+ ordered product of two operators

$$T_+(q) = \int d^4x \, e^{iqx} \langle \alpha | T_+ O_1(x) \, O_2(0) | \beta \rangle. \qquad \text{(II-44)}$$

An integration by parts gives

$$T_+(q) = (i/q^-) \int d^4x \, e^{iqx} \partial_+ \langle \alpha | T_+ O_1(x) \, O_2(0) | \beta \rangle$$

$$= (i/q^-) \int d^2 x_\perp dx^- \, e^{iq^+ x^-} e^{-iq_\perp \cdot x_\perp} \langle \alpha | [O_1(x), O_2(0)] | \beta \rangle_{x^+ = 0}$$

$$+ O[(1/q^-)^2]. \qquad \text{(II-45)}$$

Not unexpectedly, the $1/q^-$ term in a x^+ order product is the light-cone commutator, just as the $1/q^0$ term in a time ordered product is the equal time commutator. The non-trivial aspect of the theorem emerges when we recall that a covariant Green's function (T^* product) is equal to the T_+ product, apart from seagulls. Consequently we have

$$T^*(q) \equiv \int d^4x \, e^{iqx} \langle \alpha | T^* O_1(x) \, O_2(0) | \beta \rangle$$

$$\xrightarrow{q^- \to \infty} \text{polynomials} + (i/q^-) \int d^2 x_\perp dx^- \, e^{iq^+ x^-} e^{-iq_\perp \cdot x_\perp} \qquad \text{(II-46)}$$

$$\cdot \langle \alpha | [O_1(x), O_2(0)] | \beta \rangle_{x^+ = 0} + \cdots.$$

A more careful analysis indicates that q^- should become large away from the real axis. In the usual way, if the limit diverges, we interpret this as the statement that the commutator has divergent matrix elements.

As an example, consider the free propagator of Boson fields. For large q^- it goes as $i/2q^+ q^-$ and $1/2q^+$ is indeed the Fournier transform of the canonical (free field) light cone commutator:

$$-\tfrac{1}{4} i \varepsilon(x^-) \, \delta^2(x_\perp).$$

III. Applications to Deep Inelastic Processes

The theory which we have developed will now be applied to the electro-production processes studied by the MIT-SLAC experiment. As is well known, that investigation gives an experimental determination of the quantity $\int d^4x \, e^{iqx} \langle p | [J^\mu(x), J^\nu(0)] | p \rangle$, where $|p\rangle$ is the spin-averaged nucleon target state, and J^μ is the electro-magnetic current. (In the next section, we generalize the present results to include spin and internal symmetry.)

A. Scaling Representation

One of the remarkable experimental observations is that the invariant functions which determine the commutator defined above have a convergent limit in the "deep inelastic region", q^2 and $v = p \cdot q$ both large, and $\omega = -q^2/2v$ fixed. We may inquire what must be the form of $\langle p| \, [J^\mu(x), J^\nu(0)] \, |p\rangle$ in *position* space, such that its Fourier transform exhibit the properly convergent form in the deep inelastic region. The desired form, which incorporates this regularity, as well as causality and current conservation is the following [20]:

$$i\langle p| \, [J^\mu(x), J^\nu(0)] \, |p\rangle$$

$$= [g^{\mu\nu}\Box - \partial^\mu \partial^\nu] \, \varepsilon(x^0) \, [\delta(x^2) \, (1/8\pi^2) \int_{-1}^{+1} d\omega \cdot \omega^{-2} \cdot \cos\omega x p \, F_L(\omega)$$

$$+ \theta(x^2) f_1(x^2, x \cdot p)] \tag{III-1}$$

$$+ [p^\mu p^\nu \Box - p \cdot \partial(\partial^\mu p^\nu + \partial^\nu p^\mu) + g^{\mu\nu}(p \cdot \partial)^2]$$

$$\varepsilon(x^0)\theta(x^2)[(1/8\pi^2) \int_{-1}^{+1} d\omega(\omega x p)^{-1} \sin\omega x \cdot p \cdot F_2(\omega) + f_2(x^2, x \cdot p)],$$

$$x^2 f_1(x^2, x \cdot p) \xrightarrow[x^2 \to 0]{} 0, \tag{III-2a}$$

$$f_2(x^2, x \cdot p) \xrightarrow[x^2 \to 0]{} 0. \tag{III-2b}$$

The non-trivial aspect of this representation is that the degree of singularity at $x^2 = 0$ is limited, and the most singular terms at that point are parametrized by the deep inelastic structure functions of Bjorken [9], $F_L(\omega)$ and $F_2(\omega)$. The $f_i(x^2, x \cdot p)$ determine the subdominant light cone singularity, as is seen from (III-2). The degree of singularity at $x^2 = 0$ exhibited by (III-1), is no worse than in free-field theory. Also the singularity is such that the light cone commutator, i.e. the restriction to $x^+ = 0$, exists.

B. Equal Time Sum Rules

The representation (III-1) is especially convenient for deriving the usual equal-time sum rules appropriate to deep-inelastic processes. Thus by going to $x^0 = 0$ in the $0i$ components of (III-1) we obtain the Schwinger-term sum rule [20].

$$\langle p| \, [J^0(x), J^i(0)] \, |p\rangle|_{x^0 = 0}$$
$$= (i/4\pi) \, \partial^i \delta^3(x) \int_{-1}^{+1} d\omega \cdot \omega^{-2} \cdot F_L(\omega). \tag{III-3}$$

Similarly by differentiating the ij components with respect to x^0, the Callan-Gross [21] sum rule emerges, at $x^0 = 0$. We have after some manipulation

$$\lim_{p_0 \to \infty} (p_0^2)^{-1} \int d^3 x \langle p | [\partial_0 J^i(x), J^k(0)] | p \rangle |_{x^0 = 0}$$

$$= \int_{-1}^{+1} d\omega [F_L(\omega) \delta^{jk} - F_2(\omega)(\delta^{jk} - \hat{p}^j \hat{p}^k)].$$

(III-4)

Clearly this procedure may be continued to obtain relations between $\int_{-1}^{+1} d\omega \cdot \omega^{2n} F_i(\omega) (i = L, 2)$ and

$$\lim_{p_0 \to \infty} (p_0^2)^{-n-1} \int d^3 x \langle p | [\partial_0^{2n+1} J^j(x), J^k(0)] | p \rangle |_{x^0 = 0}.$$

One disadvantage of these relations is that one needs to compute more and more obscure commutators, which depend on dynamics in a complicated way. Furthermore, it would be preferable to have *one* relation determining $F_i(\omega)$, rather than the *infinite* number of moment relations for $\int d\omega \, \omega^{2n} F_i(\omega)$. The light-cone commutators provide these desirable results.

C. Light Cone Sum-Rules

Rather than restricting the representation (III-1) to equal times, $x^0 = 0$, we may just as well restrict it to the light cone, $x^+ = 0$. It is easy to show that

$$\langle p | [J^+(x), J^+(0)] | p \rangle |_{x^+ = 0}$$

$$= \partial_- \partial_- \left\{ (i/16\pi) \varepsilon(x^-) \delta^2(x_\perp) \int_{-1}^{1} d\omega \cdot \omega^{-2} \cdot \cos\omega x^- p^+ F_L(\omega) \right\}.$$

(III-5)

Comparing this with the model for the light-cone commutator, given in (II-31) or (II-39), we have [22]

$$(4\pi)^{-1} \int_{-1}^{+1} d\omega \cdot \omega^{-2} \cos\omega x^- p^+ F_L(\omega) = \langle p | S | p \rangle = 0.$$

(III-6)

The last equality follows from the fact that the Schwinger term is taken to be a c number; hence its connected matrix element vanishes.

Eq. (III-6) contains the Schwinger term sum rule, (III-3); the Callan-Gross sum rule for F_L (III-4); as well as *all* the further moment sum rules for $\int_{-1}^{1} d\omega \, \omega^{2n} F_L(\omega)$. It is very important to note that (III-6) may be inverted to obtain the *unintegrated* result

$$F_L(\omega) = 0.$$

(III-7)

Thus Eq. (III-7) may be arrived at *without* use of positivity for $F_L(\omega)$. This happens for all the deep inelastic current algebra sum rules: relations which in that context involved an integration over ω, are now re-derived in an unintegrated form.

Further results are obtained from the $+-$ components. For simplicity we set F_L to zero, as indicated by (III-7). It then follows from (III-1) that

$$\langle p|\,[J^+(x), J^-(0)]\,|p\rangle|_{x^+=0}$$

$$= -i\varepsilon(x^-)\,\delta^2(\boldsymbol{x}_\perp)\left[f(x^-p^+) + (p_\perp^2/8\pi)\int_{-1}^{1} d\omega\,\cos\omega x^-p^+F_2(\omega)\right] \qquad \text{(III-8)}$$

$$- i\varepsilon(x^-)\,p^i\partial_i\delta^2(\boldsymbol{x}_\perp)\,(1/8\pi)\int_{-1}^{1} d\omega\,\sin\omega x^-p^+F_2(\omega)/\omega\,.$$

Here $f(\alpha)$ parametrizes corrections to scaling, and will not concern us here; but see Chapter IV.

$$f(\alpha) = (M^2/8\pi)\int_{-1}^{1} d\omega\,\cos\omega\alpha F_2(\omega) + \partial/\partial\alpha\,[\alpha f_1(0, \alpha)]\,. \qquad \text{(III-9)}$$

The expression (III-8) may be compared with the operator formula (we ignore c number Schwinger terms)

$$[J^+(x), J^-(y)]_{x^+=y^+}$$

$$= -\tfrac{1}{2}\,\partial_-^x\,\{\varepsilon(x^- - y^-)\,\delta^2(\boldsymbol{x}_\perp - \boldsymbol{y}_\perp)\,V^-(x|y)\} \qquad \text{(III-10)}$$

$$- \tfrac{1}{4}\,\partial_i^x\,\{\varepsilon(x^- - y^-)\,\delta^2(\boldsymbol{x}_\perp - \boldsymbol{y}_\perp)\,[V^i(x|y) + i\varepsilon^{ij}A_j(x|y)]\} - \text{h.c.}$$

In a Fermion model, from which (III-10) is abstracted, we have

$$J^\mu(x) = \bar{\psi}(x)\,\gamma^\mu Q\,\psi(x);\quad V^\mu(x|y) = \bar{\psi}(x)\,\gamma^\mu Q^2\psi(y)\,,$$

$$A^\mu(x|y) = i\bar{\psi}(x)\,\gamma^\mu\gamma_5 Q^2\psi(y)\,,$$

where Q is the charge matrix. (In the *quark* model a more specific formula may be obtained since Q is given by the $SU(3)$ λ matrices.) Consequently (III-6) and (III-10) imply [22]

$$-(p^i/2\pi)\int_{-1}^{1} d\omega\,\sin\omega x^-p^+F_2(\omega)/\omega = \langle p|\,iV^i(x|0) + \text{h.c.}\,|p\rangle_{x^2=0}$$

$$\qquad \text{(III-11)}$$

$$= \langle p|\,i\bar{\psi}(x)\,\gamma^i Q^2\psi(0) + \text{h.c.}\,|p\rangle|_{x^2=0}\,.$$

As always, the bilocal operator is evaluated on the light cone, $x^2 = 0$, with $x^+ = 0$, $\boldsymbol{x}_\perp = 0$.

This then is the final, remarkably simple result, which summarizes *all* the moment sum rules. It is very gratifying that the bilocal operator, whose matrix elements determine $F_2(\omega)$ has no dependence on the gluon field (in the gauge $B^+ = 0$). As was noted before, the bilocal current is a simple generalization of the usual local current. Now we see that just as the matrix elements of the *local* current (the form factors) are relevant to *elastic* scattering, so analogously the matrix elements of the *bilocal* current determine the form factors appropriate to *deep inelastic* scattering. Of course we do not have at the present time an evaluation of

$$\langle p|\, i\,\overline{\psi}(x)\, Q^2\gamma^\mu\psi(0) + \text{h.c.}\, |p\rangle\,,$$

just as at the present time one cannot evaluate the elastic form factors $\langle p'|\overline{\psi}(0)\, Q\gamma^\mu\psi(0)\,|p\rangle$. The importance of the result (III-11) lies in the fact that it demonstrates the measurability of the matrix elements of bilocal operators. Thus these operators should be considered on the same footing as local currents in their relevance for practical physics.

Eq. (III-11) may be considered as a starting point for the calculation of the deep inelastic structure functions. Rather than computing the cross-sections for all q^2 and v, and then passing to the deep inelastic limit, one needs to compute only the matrix elements in (III-11).

It is clear that similar investigations can be carried out for the deep inelastic neutrino processes. All previous current algebraic sum rules which provided relations between *moments* of the appropriate structure functions are now replaced by relations between *Fourier* transforms of these functions; or equivalently between the functions themselves.

We conclude this section with several further (unrelated) observations about the bilocal operators. If $i\,\overline{\psi}(x)\,\gamma^\mu\psi(0)|_{x^2=0}$ is expanded at small x^- in powers of x^-, one encounters local structures of the form

$$\overline{\psi}\gamma^\mu\overleftrightarrow{\partial}^{v_1}\dots\overleftrightarrow{\partial}^{v_n}\psi\,.$$

In particular, the *Fermion* part of the energy-momentum tensor occurs in this expansion – though the *Boson* part is missing [22]. For large x^-, only small ω, i.e. the Regge region, contributes. Thus the bilocal operator at large x^- parametrizes Regge poles. Finally we repeat once more the *caveat* that in perturbation theory the commutators possess anomalies and are not of the canonical form given here. One consequence of this is that $F_L \neq 0$, in perturbation theory, contrary to (III-7) [23]. Evidently we are ignoring here these perturbative q number anomalies.

IV. Fixed-Mass Sum Rules

As Prof. Furlan has explained in his lectures here [3], the ordinary current algebra is insufficient to derive fixed mass sum rules, like the one of Dashen, Fubini and Gell-Mann [1, 2]. One possible technique which *does* yield fixed mass sum rules, is the $p \to \infty$ procedure. Unfortunately that method frequently fails because of illegitimate interchange of limit and integral and many of the results are invalid even in free-field models. On the other hand it has been known for some time that the fixed mass sum rules are equivalent to appropriate light cone commutators and that the $p \to \infty$ technique is an attempt (not always successful) to determine the (then unknown) light cone commutator from the (then known) equal time commutator [4]. However since we now have a model for the light cone commutator we can dispense with the unreliable $p \to \infty$ method. We derive in this section corrected versions of various fixed mass sum rules [17].

A. Kinematic Preliminaries

We consider the diagonal matrix element of the commutator of two vector currents V_a^μ, assumed to be conserved, between Fermion states, which are *not* spin averaged.

$$C_{ab}^{\mu\nu}(p, q) = \int d^4 x \, e^{iqx} \langle p | [V_a^\mu(x), V_b^\nu(0)] | p \rangle . \qquad (\text{IV-1})$$

The Fermion state $|p\rangle$ is characterized by a spin (pseudo) vector $s^\alpha \equiv \bar{u}(p) \, i \gamma^\alpha \gamma_5 u(p)$. This vector is orthogonal to p and has the form $s^\mu = (s^0, \vec{s}), s^0 = \vec{p} \cdot \hat{n}, \vec{s} = m\hat{n} + \vec{p}\vec{p} \cdot \hat{n}/[E+m]$. Here $\hat{n}$ is an arbitrary unit vector specifying the rest frame spin direction, $\hat{n} = \langle \vec{\sigma} \rangle$. Note that $p^2 = -s^2 = m^2$.

Considerations of parity, time inversion invariance, Lorentz invariance and current conservation give the following expression for $C_{ab}^{\mu\nu}(p, q)$.

$$\begin{aligned} C_{ab}^{\mu\nu}(p, q) = & [-g^{\mu\nu} + q^\mu q^\nu/q^2] \, W_L^{ab} \\ & + [p^\mu p^\nu - (v/q^2)(p^\mu q^\nu + p^\nu q^\mu (+ g^{\mu\nu} v^2/q^2] \, W_2^{ab} \\ & + i\varepsilon^{\mu\nu\alpha\beta} s_\alpha q_\beta W_3^{ab} + i\varepsilon^{\mu\nu\alpha\beta} p_\alpha q_\beta q \cdot s \, W_4^{ab} . \end{aligned} \qquad (\text{IV-2})$$

The invariants are functions of q^2, v. We shall decompose them into parts symmetric in ab, denoted by (ab); and antisymmetric in ab, denoted by $[ab]$:

$$W_i^{ab} = W_i^{(ab)} + i W_i^{[ab]}, \quad i = L, 2, 3, 4 . \qquad (\text{IV-3})$$

Crossing now implies that

$$W_i^{(ab)}(q^2, v) = - W_i^{(ab)}(q^2, -v). \quad i = L, 2, 3, \qquad (\text{IV-4a})$$

$$W_4^{(ab)}(q^2, v) = W_4^{(ab)}(q^2, -v) \qquad (\text{IV-4b})$$

and the opposite symmetry obtains for the invariants which are anti-symmetric in a and b. Hermiticity insures that the invariants occurring in the right hand side of (IV-3) are real.

Since the current has dimension 3, and since our states are covariantly normalized, the following functions are dimensionless: W_L^{ab}, $v\,W_2^{ab}$, $v\,W_3^{ab}$ and $v^2 W_4^{ab}$. Hence these objects approach a limit as $v \to \infty$ with fixed $-q^2/2v \equiv \omega$ [3]. In the present context, this is *not* a hypothesis, but a consequence of our use of the light cone commutators. In the pre-asymptotic region, we shall frequently consider the above functions to depend on ω and q^2. In that case W_L^{ab} will be denoted by $(-1/2\omega)\tilde{F}_L^{ab}(\omega, q^2)$, and the others by $\tilde{F}_i^{ab}(\omega, q^2)$. In the scaling limit, the $\tilde{F}_i^{ab}(\omega, q^2)$ become $F_i^{ab}(\omega)$, $i = L, 2, 3, 4$. The quark model commutator structure which we employ implies that $F_L^{ab}(\omega)$ vanishes, as will be seen below. We shall need the correction to scaling for this function. Hence we define $\lim\limits_{-q^2 \to \infty} q^2 \tilde{F}_L^{ab}(\omega, q^2) \equiv G_L^{ab}(\omega)$. One can show that F_i^{ab} and G_L^{ab} vanish for $|\omega| > 1$.

B. Infinite Momentum Derivation of Sum Rules

We summarize the $p \to \infty$ sum rules which emerge from the $0v$ components of (IV-1). Some of these results are well known, others presumably have been recorded in the literature. Eq. (IV-1) and the usual current algebra imply

$$(2\pi)^{-1} \int_{-\infty}^{\infty} dq^0 C_{ab}^{0\alpha}(p, q) = i f_{abc} p^\alpha \Gamma_c$$

$$\langle p| V_a^\alpha(0) |p\rangle \equiv p^\alpha \Gamma_a \,. \tag{IV-5}$$

We have taken the Schwinger term to be a c number. Throughout subsection B we set $\vec{p} \cdot \vec{q} = 0$. A change of variable is now performed $q^0 = v/p^0$.

$$(2\pi)^{-1} \int_{-\infty}^{\infty} (dv/p^0)\, C_{ab}^{0\alpha}(p, q)\Big|_{\substack{q^0 = v/p^0, \\ \vec{p} \cdot \vec{q} = 0}} = i f_{abc} p^\alpha \Gamma_c \,. \tag{IV-6}$$

The above is exact, however it is not a *fixed mass* sum rule since $q^2 = v^2/p_0^2 - q^2$ varies with v. To obtain fixed mass sum rules, p_0 is set to ∞ in the usual fashion. Then by equating independent tensors in (IV-6b), the following non-trivial relations are obtained.

From $\alpha = 0$ in (IV-6), the Dashen, Fubini, Gell-Mann [1, 2] result emerges.

$$\int_0^\infty dv\, W_2^{[ab]}(q^2, v) = \pi f_{abc} \Gamma_c \qquad q^2 \leq 0 \,. \tag{IV-7a}$$

---

[3] This is in complete analogy with the electroproduction results discussed above.

From $\alpha = i$ in (IV-6) the "good-bad" sum rules are a consequence

$$\int_0^\infty d\nu\, W_3^{[ab]}(q^2, \nu) = 0, \tag{IV-7b}$$

$$\int_0^\infty d\nu\, \nu\, W_4^{[ab]}(q^2, \nu) = 0, \tag{IV-7c}$$

$$\int_0^\infty d\nu\, W_L^{[ab]}(q^2, \nu) = 0, \tag{IV-7d}$$

$$\int_0^\infty d\nu\, W_4^{(ab)}(q^2, \nu) = 0, \tag{IV-7e}$$

$$\int_0^\infty d\nu\, \nu\, W_2^{(ab)}(q^2, \nu) = 0 \qquad q^2 \leqq 0. \tag{IV-7f}$$

The sum of (IV-7b) and (IV-7c) is the Bég sum rule [24], which when evaluated at $q^2 = 0$ gives a relation for anomalous magnetic moments.

If the results (IV-7) are compared with the explicit calculations in the *free* quark model, it is then found that (IV-7b) and (IV-7f) are not valid, while (IV-7c), (IV-7d) and (IV-7e) are trivially satisfied, in the sense that both sides of the equality vanish. Numerical evaluation [25], which checks the validity of these in nature is consistent with (IV-7a), in that the Cabbibo-Radicatti relation [26] which is a consequence of (IV-7a), appears to be satisfied. However the Bég sum rule [24] is not verified experimentally. Finally (IV-7f) cannot be true since it can be rewritten in terms of the scaling variables as $\int_{-1}^{1} d\omega \cdot \omega^{-2} \cdot \tilde{F}_2^{(ab)}(\omega, q^2) = 0$. When q^2 is then set to ∞, one gets $\int_{-1}^{1} d\omega \cdot \omega^{-2} \cdot F_2^{(ab)}(\omega) = 0$ which is inconsistent with the MIT-SLAC experiments which show that $F_2^{(ab)}(\omega) \neq 0$ [5].

In the free-field model all integrals converge, since the invariants are δ functions. Hence the failures in that context cannot be attributed to divergences. In a Regge pole model the relation which diverge are (IV-7d) and (IV-7f). Note that the Bég sum rule converges in a Regge pole model [27].

C. Light Cone Derivation of Sum Rules

To derive the fixed mass sum rules with light cone techniques, we set q^+ to zero in the $+\nu$ components of (IV-1), and integrate over q^-.

$$(2\pi)^{-1} \int_{-\infty}^{\infty} dq^- \, C_{ab}^{+\alpha}(p, q)|_{q^+ = 0}$$
$$= \int d^2 x_\perp e^{-i q_\perp \cdot x_\perp} \langle p| \left[\int dx^- V_a^+(x), V_b^\alpha(0) \right] |p\rangle|_{x^+ = 0}. \tag{IV-8}$$

With this procedure, we obtain directly a fixed mass sum rule, since $q^2 = -q_\perp^2$ when $q^+ = 0$; i.e. q^2 does not vary in the course of the integration. However, the right hand side of (IV-8) involves a light cone commutator, rather than an equal time commutator. We assume that this object can be computed from our Ansatz, (II-39), (II-40), and (II-41).

$$\langle p|\, [\int dx^- V_a^+(x),\, V_b^\alpha(0)]\, |p\rangle|_{x^+=0}$$
$$= \int dx^- \langle p|\, [V_a^+(x),\, V_b^\alpha(0)]\, |p\rangle|_{x^+=0}. \qquad \text{(IV-9)}$$

The nature of this assumption shall be discussed further in the Conclusion. Finally we change variables in (IV-8) by setting $v = p^+ q^- - p_\perp \cdot q_\perp$. This yields

$$(2\pi)^{-1} \int_{-\infty}^{\infty} (dv/p^+)\, C_{ab}^{+\,\alpha}(p,q)\Big|_{\substack{q^+=0 \\ q^-=(v+p_\perp \cdot q_\perp)/p^+}}$$
$$= \int d^2x_\perp dx^-\, e^{-iq_\perp \cdot x_\perp} \langle p|\, [V_a^+(x),\, V_b^\alpha(0)]\, |p\rangle|_{x^+=0}. \qquad \text{(IV-10)}$$

We now use the model given previously to evaluate the commutator in (IV-10) for $\alpha = +, -$. (No new information emerges from $\alpha = i$). The Schwinger terms are ignored as they are taken to be c numbers. Also since the current is conserved, the commutators do not have symmetry breaking terms in them. Clearly the right hand side of (IV-10) will involve matrix elements of the bilocal operators. These can be given by the following form factor decomposition. Define

$$\mathscr{V}_a^\mu(x|y) \equiv \tfrac{1}{2} V_a^\mu(x|y) + \tfrac{1}{2} V_a^\mu(y|x), \qquad \text{(IV-11a)}$$

$$\bar{\mathscr{V}}_a^\mu(x|y) \equiv \tfrac{1}{2i} V_a^\mu(x|y) - \tfrac{1}{2i} V_a^\mu(y|x), \qquad \text{(IV-11b)}$$

$$\mathscr{A}_a^\mu(x|y) \equiv \tfrac{1}{2} A_a^\mu(x|y) + \tfrac{1}{2} A_a^\mu(y|x), \qquad \text{(IV-11c)}$$

$$\bar{\mathscr{A}}_a^\mu(x|y) \equiv \tfrac{1}{2i} A_a^\mu(x|y) - \tfrac{1}{2i} A_a^\mu(y|x). \qquad \text{(IV-11d)}$$

The form factors are by definition

$$\langle p|\, \mathscr{V}_a^\mu(x|0)\, |p\rangle = p^\mu V_a^1(x^2, x\cdot p)$$
$$+ x^\mu V_a^2(x^2, x\cdot p), \qquad \text{(IV-12a)}$$

$$\langle p|\, \bar{\mathscr{V}}_a^\mu(x|0)\, |p\rangle = p^\mu \bar{V}_a^1(x^2, x\cdot p)$$
$$+ x^\mu \bar{V}_a^2(x^2, x\cdot p), \qquad \text{(IV-12b)}$$

$$\langle p|\, \mathscr{A}_a^\mu(x|0)\, |p\rangle = s^\mu A_a^1(x^2, x\cdot p)$$
$$+ p^\mu x\cdot s\, A_a^2(x^2, x\cdot p) \qquad \text{(IV-12c)}$$
$$+ x^\mu x\cdot s\, A_a^3(x^2, x\cdot p),$$

$$\langle p|\, \bar{\mathscr{A}}_a^\mu(x|0)\, |p\rangle = s^\mu \bar{A}_a^1(x^2, x\cdot p)$$
$$+ p^\mu x\cdot s\, \bar{A}_a^2(x^2, x\cdot p) \qquad \text{(IV-12d)}$$
$$+ x^\mu x\cdot s\, \bar{A}_a^3(x^2, x\cdot p).$$

T inversion invariance eliminates a possible structure of the form $\varepsilon^{\mu\alpha\beta\varrho} x_\alpha p_\beta s_\varrho$ in (IV-12a) and (IV-12b). It is clear that (IV-12a) and (IV-12c) are even in x, while (IV-12b) and (IV-12d) are odd in x.

It is now straightforward to extract the sum rules by equating independent tensors in (IV-10). The results are [17]

$$\int_0^\infty d\nu\, W_2^{[a,b]}(q^2, \nu) = \pi f_{abc} \Gamma_c\,, \tag{IV-13a}$$

$$\int_0^\infty d\nu\, W_3^{[ab]}(q^2, \nu) = \tfrac{1}{2}\pi f_{abc} \int_0^\infty d\alpha\, \bar{A}_c^1(0, \alpha)\,, \tag{IV-13b}$$

$$\int_0^\infty d\nu\, \nu\, W_4^{[ab]}(q^2, \nu) = \tfrac{1}{2}\,\pi f_{abc} \int_0^\infty d\alpha\, \bar{A}_c^2(0, \alpha)\,, \tag{IV-13c}$$

$$\int_0^\infty d\nu\, W_L^{[ab]}(q^2, \nu) = 0\,, \tag{IV-13d}$$

$$\int_0^\infty d\nu\, W_4^{(ab)}(q^2, \nu) = 0\,, \tag{IV-13e}$$

$$\int_0^\infty d\nu (\nu/-q^2)\, W_2^{(ab)}(q^2, \nu) = \tfrac{1}{2}\,\pi d_{abc} \int_0^\infty d\alpha\, \bar{V}_c^1(0, \alpha)\,, \qquad q^2 \leqq 0. \tag{IV-13f}$$

The Dashen, Fubini, Gell-Mann sum rule [1, 2] (IV-13a), as well as (IV-13d), and (IV-13e) are rederived; but the Bég sum rules [24] (IV-13b) and (IV-13c) as well as (IV-13f) are found to have corrections. This is extremely gratifying, since it is precisely these which fail both in free-field theory and in nature [25]. The corrections are expressed in terms of integrals over matrix elements of the bilocal operators. In the next subsection we shall show that these matrix elements are measurable, even when an integration is not performed. Here we merely demonstrate how (IV-13b), (IV-13c) and (IV-13f) can be exploited.

Observe that the right hand sides of (IV-13b), (IV-13c) and (IV-13f) are independent of q^2. Let us rewrite the left hand sides in terms of the scaling functions.

$$\int_0^\infty d\nu\, W_3^{[ab]}(q^2, \nu) = \int_0^1 d\omega\, \omega^{-1}\, \tilde{F}_3^{[ab]}(\omega, q^2)\,, \tag{IV-14a}$$

$$\int_0^\infty d\nu\, \nu\, W_4^{[ab]}(q^2, \nu) = \int_0^1 d\omega\, \omega^{-1}\, \tilde{F}_4^{[ab]}(\omega, q^2)\,, \tag{IV-14b}$$

$$\int_0^\infty d\nu (\nu/-q^2)\, W_2^{(ab)}(q^2, \nu) = \int_0^1 d\omega (2\omega^2)^{-1}\, \tilde{F}_2^{(ab)}(\omega, q^2), \qquad q^2 \leqq 0. \tag{IV-14c}$$

According to the sum rules, the integrals are q^2 independent, and may be evaluated by letting $-q^2 \to \infty$. Thus the right hand sides of (IV-14) are expressible in terms of the deep inelastic structure functions $F_i^{ab}(\omega)$. Performing a similar analysis of the remaining sum rules in (IV-13) gives finally the following results [17].

$$\int_0^\infty dv \, W_2^{[ab]}(q^2, v) = \int_0^1 d\omega \, \omega^{-1} F_2^{[ab]}(\omega) = \pi f_{abc} \Gamma_c \, , \qquad \text{(IV-15a)}$$

$$\int_0^\infty dv \, W_3^{[ab]}(q^2, v) = \int_0^1 d\omega \, \omega^{-1} F_3^{[ab]}(\omega) \, , \qquad \text{(IV-15b)}$$

$$\int_0^\infty dv \, v \, W_4^{[ab]}(q^2, v) = \int_0^1 d\omega \, \omega^{-1} F_4^{[ab]}(\omega) \, , \qquad \text{(IV-15c)}$$

$$\int_0^\infty dv \, W_L^{[ab]}(q^2, v) = \int_0^1 d\omega \, \omega^{-3} F_L^{[ab]}(\omega) \, , \qquad \text{(IV-15d)}$$

$$\int_0^\infty dv \, W_4^{(ab)}(q^2, v) = \int_0^1 d\omega \, F_4^{(ab)}(\omega) \, , \qquad \text{(IV-15e)}$$

$$\int_0^\infty dv \, v(-q^2)^{-1} W_2^{(ab)}(q^2, v) = \int_0^1 d\omega \, (2\omega^2)^{-1} F_2^{(ab)}(\omega) \, . \qquad \text{(IV-15f)}$$

The sum rule (IV-15f) has already been derived by an entirely different method, by Cornwall, Corrigan and Norton [28]. Note that this relation diverges in a Regge pole model. Cornwall, Corrigan and Norton discuss a truncation technique, which possibly might give meaning to this divergent sum rule.

D. Measuring Bilocal Operators

It was seen in our discussion of deep inelastic scattering that matrix elements of the bilocal operators, with light-like separated arguments are completely measurable in terms of deep inelastic structure functions. Clearly the same is true here. This can be established in one of two ways. In complete analogy to the discussion of section IV, one may exhibit a position space representation for $(2\pi)^{-1} \int d^4q \, e^{-iqx} C_{ab}^{\mu\nu}(p, q)$ which incorporates scaling as in (III-1). Alternatively one may work directly in momentum space and use the light-cone version of the BJL limit discussed in Section II. The results are as follows [17].

$$d_{abc} \bar{V}_c^1(0, \alpha) + f_{abc} V_c^1(0, \alpha) = (2\pi i)^{-1} \int_{-1}^1 d\omega \, \omega^{-1} e^{i\omega\alpha} F_2^{ab}(\omega) \, , \qquad \text{(IV-16a)}$$

$$d_{abc} A_c^1(0, \alpha) - f_{abc} \bar{A}_c^1(0, \alpha) = \pi^{-1} \int_{-1}^1 d\omega \, e^{i\omega\alpha} F_3^{ab}(\omega) \, , \qquad \text{(IV-16b)}$$

$$d_{abc} \alpha \, A_c^2(0, \alpha) - f_{abc} \alpha \, \bar{A}_c^2(0, \alpha) = \pi^{-1} \int_{-1}^1 d\omega \, e^{i\omega\alpha} F_4^{ab}(\omega) \, , \qquad \text{(IV-16c)}$$

$$d_{abc} \alpha \, \bar{V}_c^2(0, \alpha) + f_{abc} \alpha \, V_c^2(0, \alpha) = \tfrac{1}{16} i \pi^{-1} \int_{-1}^1 d\omega \, \omega^{-3} e^{i\omega\alpha} G_L^{ab}(\omega) \, . \qquad \text{(IV-16d)}$$

This is the generalization to spin and internal symmetry of (III-11). Also one finds [17]

$$F_L^{ab}(\omega) = 0 \qquad \qquad (\text{IV-17})$$

which is the generalization of (III-7). (Here $F_i^{ab} = F_i^{(ab)} + i F_i^{[ab]}$).

Additional results follow from (IV-16). Setting α to zero in (IV-16a) we have, since $\bar{V}_c^1(0, \alpha)$ is odd in α,

$$f_{abc} V_c^1(0, 0) = \pi^{-1} \int_0^1 d\omega \, \omega^{-1} F_2^{[ab]}(\omega). \qquad (\text{IV-18a})$$

The right hand side can be evaluated from the Dashen, Fubini, Gell-Mann [1, 2] sum rule, evaluated in the scaling region, (IV-15a). Thus (IV-18a) becomes

$$f_{abc} V_c^1(0, 0) = f_{abc} \Gamma_c. \qquad (\text{IV-18b})$$

Let us recall now the definitions of $V_c^1(0, 0)$ and Γ_c.

$$p^\mu f_{abc} V_c^1(0, 0) = f_{abc} \langle p | \mathcal{V}_c^\mu(0|0) | p \rangle$$
$$= f_{abc} \langle p | V_c^\mu(0|0) | p \rangle, \qquad (\text{IV-19})$$

$$p^\mu \Gamma_c = \langle p | V_c^\mu(0) | p \rangle. \qquad (\text{IV-20})$$

Thus (IV-18) shows that the proton matrix element of the bilocal operator $f_{abc} V_c^\mu(x|y)$ reduces to that of the vector current as $x \to y$. This of course is obvious in the field theoretic model considered where $V_c^\mu(x|y) = \bar{\psi}(x) \frac{1}{2} \lambda_c \gamma^\mu \psi(y)$. However, if we generalize and abstract from the model, we see that (IV-18) places an important model independent constraint on the bilocal operator $V_c^\mu(x|y)$.

We now assume that the same is true for the axial vector bilocal operator, as is of course the case in our model. We have from [17] (IV-16b)

$$d_{abc} A_c^1(0, 0) s^\mu = d_{abc} \langle p | A_c^\mu(0|0) | p \rangle$$
$$= d_{abc} \langle p | A_c^\mu(0) | p \rangle = d_{abc} \Gamma_c^A s^\mu \qquad (\text{IV-21})$$
$$= 2 s^\mu \pi^{-1} \int_0^1 d\omega \, F_3^{(ab)}(\omega)$$

($\bar{A}_c^1(0, 0)$ vanishes by symmetry.) The sum rule which has emerged, relating Γ_c^A to the spin odd, isospin even deep inelastic cross section is similar to a relation first obtained in the quark model by Bjorken [29, 30].

V. Conclusion

It is clear that the approach which we have here developed yields very informative and elegant results. A question which naturally arises is how generally valid are the conclusions of our investigation. It is this topic which we shall discuss in this Section. We shall not elaborate again on the canonical, formal nature of the present results, which invalidates them in perturbation theory.

The bilocal operators which occur in the light cone commutators are certainly model dependent, even when they are interaction independent as in (II-39), (II-40) and (II-41). As we have already mentioned, scalar electrodynamics gives entirely different expressions, and the same is true of a Yang-Mills theory. These various commutator models may be said to differ among themselves in the space-time tensor structure. In addition to this there is the internal symmetry model dependence, which we have taken to be the $SU(3)$ triplets of the quark model.

It is plausible to suppose that nature favors the space-time tensor structure of the Fermion model, since the other models have non-vanishing F_L or $F_2 = F_L$ [17, 20, 21]. In this connection it would be most interesting to explore what *model independent* information can be obtained about the $[V_a^+, V_b^\alpha]$ light cone commutator, if it is assumed that for $\alpha = +$ the form is as in (II-39)[4]. The internal symmetry structure which we have used does not, at the present time, possess any convincing experimental verification. Our results are true only in the triplet realization of $SU(3)$ where $\lambda_a\lambda_b = (i f_{abc} + d_{abc}) \lambda_c$. To generalize beyond this, one could for example set $\lambda_a\lambda_b = i f_{abc}\lambda_c + d_{(ab)}$ where $d_{(ab)}$ is no longer $d_{abc}\lambda_c$, but contains all the symmetric parts of the $8 \otimes 8$ representation. Even further, one might generalize by setting $\lambda_a\lambda_b = i f_{[ab]} + d_{(ab)}$ where $f_{[ab]}$ is no longer the pure octet $f_{abc}\lambda_c$ but rather is a general anti-symmetric $SU(3)$ matrix.

With these *internal symmetry* generalization, the commutators would retain the same form as before, except that the $SU(3)$ content of the bilocal operators would be more complicated. One would need to replace $d_{abc}V_c^\mu(x|y)$ and $d_{abc}A_c^\mu(x|y)$ by $V_{(ab)}^\mu(x|y)$ and $A_{(ab)}^\mu(x|y)$, and similarly for the antisymmetric combinations. Consequently (IV-16) would possess left hand sides with more complicated $SU(3)$ structures. Thus the measurement of the internal symmetry of the deep inelastic structure functions will provide important information concerning the validity of specific models. For example the inequality of the proton and neutron

---

[4] This problem is analogous to the model-independent determination of the $[V_a^0, V_b^i]$ equal-time commutator from the $[V_a^0, V_b^0]$ equal-time commutator [31].

data [5] shows that the bilocal operators are not $SU(2)$ singlets. In this context, we see that (IV-18) is an important, model independent constraint.

The corrected deep inelastic sum rules (IV-15b), (IV-15c) and (IV-15f), as well as the first equality in (IV-15a), (IV-15d) and (IV-15e) are quite model independent, since they are consequences merely of (IV-9), causality and scaling. The point is that according to (IV-8) and (IV-9) a fixed mass sum rule is given by an integral over x^- and a Fourier transform with respect to $q_\perp$ of a light-cone commutator. This commutator is local in $x_\perp$ due to causality, i.e. it is composed of δ functions of $x_\perp$ and derivatives of δ functions. Consequently in Fourier space the integral over v of an invariant function $W(-q_\perp^2, v)$ must be a polynomial in $q_\perp^2$. But the degree of the polynomial is fixed by scaling. Specifically, for example for (IV-15b) we can conclude quite generally from (IV-9) and locality

$$\int_0^\infty dv\, W_3^{[ab]}(-q_\perp^2, v) = \int_0^1 d\omega\, \omega^{-1}\, \tilde{F}_3^{[ab]}(\omega, -q_\perp^2)$$

$$\text{(V-1a)}$$

$$= \sum_{n=0}^M (q_\perp^2)^n\, C_n^{[ab]}.$$

However scaling requires that the limit $q_\perp^2 \to \infty$ of (V-1a) exists; hence we learn that

$$\int_0^\infty dv\, W_3^{[ab]}(-q_\perp^2, v) = C_0^{[ab]} = \int_0^1 d\omega\, \omega^{-1}\, F_3^{[ab]}(\omega) \qquad \text{(V-1b)}$$

which is (IV-15b). The second equality in (IV-15a), (IV-15d) and (IV-15e) is however a consequence of the specific model considered.

Finally we come to the question of the validity of (IV-9), which is seen to involve an interchange of integration over x^- with the limit $x^+ \to 0$. A careful investigation of this has been given elsewhere [17] and the result is the following. In the terminology of Adler and Dashen [32] the class 2 graphs, i.e. the graphs which have fixed mass singularities in the current lines, must satisfy the same superconvergence relation which is required for the validity of the $p \to \infty$ technique. Thus as far as the class 2 graphs are concerned, the present considerations have nothing new to offer. On the other hand, the Z graphs, which were improperly treated by the $p \to \infty$ method, seem to be now accurately included, as is seen from the free-field example.

Put in another way, the $p \to \infty$ technique for (IV-7b) to (IV-7f) requires the absence of all fixed poles; while the present method can account for fixed poles with residues which are polynomials in q^2; see (IV-15b) to (IV-15f). If there are non-polynomial residues, then the light-cone method also fails due to the invalidity of (IV-9)[5].

References

1. Dashen, R. F., Gell-Mann, M.: Proceedings of the 3rd Coral Gables Conference on Symmetry Principles at High Energy, San Francisco: W. H. Freeman and Co.: 1966.
2. Fubini, S.: Nuovo Cimento 34A, 475 (1966).
3. Furlan, G.: this volume p. 118—147.
4. Bardakci, K., Segre, G.: Phys. Rev. 153, 1263 (1967). — Jersak, J., Stern, J.: Nuovo Cimento 59, 315 (1969). — Leutwyler, H.: Springer Tracts Mod. Phys. 50, 29 (1969).
5. For a summary see Taylor, R. E.: SLAC Report No. SLAC-PUB-677.
6. Jackiw, R., Van Royen, R., West, G. B.: Phys. Rev. D 2, 2473 (1970). — Other investigations include Ioffe, B. L.: Zh. Eksperim. Theor. Fiz. Pis'ma V Redaktsiyu 9, 163 (1969) [Sov. Phys. JETP Letters 9, 97 (1969)]. — Brown, L. S.: Lectures in Theoretical Physics, Buttin, W. E., Downs, B. W., Downs, J. (Eds.). New York: Interscience (to be published). — Brandt, R.: Phys. Rev. Letters 23, 1260 (1969). — Leutwyler, H., Stern, J.: Nucl. Phys. B 20, 77 (1970).
7. Fritsch, H., Gell-Mann, M.: Talk given at the Coral Gables Conference on Particle Physics, January 1971.
8. Gross, D. J., Treiman, S. B.: Phys. Rev. (in press).
9. That scaling should occur in the electroproduction experiments was first suggested by Bjorken, J. D.: Phys. Rev. 179, 1547 (1969).
10. The derivation is due to Cornwall, J. M., Jackiw, R.: Phys. Rev. (in press).
11. Quantization of a quantum theory on the light cone was first carried out by Kogut, J. B., Soper, D. E.: Phys. Rev. D 1, 2901 (1970).
12. Rohrlich, F.: Acta Phys. Aust. 32, 87 (1970). Examines the quantization of theories on the light cone, with particular emphasis on the associated boundary value problem.
13. Weinberg, S.: Phys. Rev. 150, 1313 (1966).
14. This argument was shown to us by Prof. S. Coleman, whom we thank.
15. This freedom also exists for massive vector meson theories, since the propagator of such particles may be chosen to be without a + component as a consequence of the conserved current interaction. For details see D. Soper, to be published.
16. Schwinger, J.: Phys. Rev. 130, 406 (1963).
17. Dicus, D., Jackiw, R., Teplitz, V.: Phys. Rev. (in press).
18. Bjorken, J. D.: Phys. Rev. 148, 1467 (1966).
19. Johnson, K., Low, F. E.: Progr. Theoret. Phys. (Kyoto) Suppl. 37—38, 74 (1966).
20. Jackiw, R., Van Royen, R., West, G. B.: Phys. Rev. D 2, 2473 (1970).
21. Callan, C. G., Jr., Gross, D. J.: Phys. Rev. Letters 22, 156 (1969).
22. This was derived by Cornwall and Jackiw, Ref. [10].
23. Jackiw, R., Preparata, G.: Phys. Rev. Letters 22, 975 (1969). — Adler, S. L., Tung, Wu-Ki: Phys. Rev. Letters 22, 978 (1969).
24. Bég, M. A. B.: Phys. Rev. Letters 17, 333 (1966).
25. Fox, G. C., Freedman, D. Z.: Phys. Rev. 182, 1628 (1969).
26. Cabibbo, N., Radicatti, L. A.: Phys. Letters 19, 697 (1966).
27. A detailed Regge analysis is given in Ref. [17].

---

[5] These considerations were developed with D. J. Gross, whose comments are much appreciated.

28. Cornwall, J. M., Corrigan, J. D., Norton, R. E.: Phys. Rev. Letters **24**, 1141 (1970) and Phys. Rev. D **3**, 536 (1971).
29. Bjorken, J. D.: Phys. Rev. **148**, 1467 (1966).
30. Gálfi, L., Gnädig, P., Kuti, J., Niedermayer, F., Patkos, A.: XV th International Conference on High Energy Physics, Kiev, Aug. 26, 1970, (to be published).
31. Gross, D. J., Jackiw, R.: Phys. Rev. **163**, 1688 (1967) and Jackiw, R.: Phys. Rev. **175**, 2015 (1968).
32. Adler, S. L., Dashen, R.: Current Algebra. New York: W. A. Benjamin 1968.

Professor Dr. Roman Jackiw
Laboratory for Nuclear Science
and
Department of Physics
Massachusetts Institute of Technology
Cambridge, Mass., USA

Local Saturation of Commutator Matrix Elements

Hans D. Dahmen

Contents

Introduction

Locality being one of the basic concepts of particle physics should not only be preserved in approximations but actually be used as a guiding principle for finding approximation schemes. It has been the objective of a number of investigations to develop local one particle approximations [1, 2] and local approximations involving the two particle structure of Greens functions [2] which go beyond simple perturbation theory.

It is the purpose of this talk to present an approach to the systematic construction of local approximations which can be completely characterized in terms of intermediate states. It could be considered a local form of a Tamm-Dancoff method [3]. The method has been developed in a number of papers in collaboration with Gromes, Rothe, and Stech [4].

1. Local Saturation

We consider a neutral scalar field $A(x)$ of mass m which is relatively local to itself, i.e.

$$[A(x), A(0)] = 0 \quad \text{for} \quad x^2 < 0. \tag{1}$$

The Haag expansion [5] of this field, i.e. the expansion of $A(x)$ in terms of Wick-monomials of "in"-fields has the form

$$\tilde{A}(k) = \tilde{A}_{\text{in}}(k) + \sum_{n=2}^{\infty} (n!)^{-1} \int \left(\prod_{i=1}^{n} d^4 k_i \right) \delta^4 \left(k - \sum_{i=1}^{n} k_i \right)$$
$$\cdot h_n(k_1, ..., k_n) : \tilde{A}_{\text{in}}(k_1) ... \tilde{A}_{\text{in}}(k_n) :, \tag{2}$$

where the Fourier transform $A(k)$ is defined by

$$A(x) = (2\pi)^{-3/2} \int d^4k\, e^{-ikx} \tilde{A}(k)\,. \tag{3}$$

The properties of the coefficient functions $h_n(k_1, \ldots, k_n)$ are the following:
1. Lorentz invariance.
2. Symmetry in the arguments.
3. Restriction of the arguments to on shell values $k_i^2 = m^2$.
4. $h_n(k_1, \ldots, k_n) = h_n^*(-k_1, \ldots, -k_n)$.
5. h_n are boundary values of analytic functions in each time component k_i^0, $(i = 1, \ldots, n)$.
6. The Fourier transforms of the h_n are retarded in all arguments.

We note that these properties do not insure the locality of $A(x)$ in the sense of Eq. (1).

Because of the properties (1–6) we make the following ansatz

$$h_2(k_1, k_2) = \frac{f_2((k_1 + k_2 + i\varepsilon)^2)}{m^2 - (k_1 + k_2 + i\varepsilon)^2}\,,$$

$$h_3(k_1, k_2, k_3) = \frac{f_3((k_1 + k_2 + i\varepsilon)^2, (k_1 + k_3 + i\varepsilon)^2, (k_1 + k_2 + k_3 + i\varepsilon)^2)}{m^2 - (k_1 + k_2 + k_3 + i\varepsilon)^2} \tag{4}$$

etc., where the $i\varepsilon$-presciption refers to the time component. The f_n are real functions of the Lorentz invariant variables

$$f_n(z_1, \ldots, z_n) = f_n^*(z_1^*, \ldots, z_n^*)\,. \tag{5}$$

The ansatz (4) assures the weak convergence of the "interpolating" field $A(x)$ to the "in"-field $A_{in}(x)$ and to the "out"-field $A_{out}(x)$ defined through

$$\tilde{A}_{out}(k) = \tilde{A}_{in}(k) + 2\pi i\, \varepsilon(k_0)\, \delta(m^2 - k^2)\, \tilde{j}(k)\,, \tag{6}$$

where the current $j(x)$ is related to $A(x)$ by

$$j(x) = (\Box + m^2)\, A(x)\,. \tag{7}$$

Again the locality of $A_{out}(x)$ relatively to itself is not guaranteed by Eq. (6).

To insure the locality of $A(x)$ as well as $A_{out}(x)$ we have to study the infinite set of equations which derives from the insertion of the Haag expansion (2) into the commutator

$$K(q) = (2\pi)^{-1} \int d^4x\, e^{iqx}[A(x), A(0)]\,. \tag{8}$$

This commutator can also be expanded into a sum of Wick-monomials of "in" fields

$$K(q) = c_0(q) + \sum_{i=1}^{n} (n!)^{-1} \int \left(\prod_{i=1}^{n} d^4k_i\right) c_n(q; k_1, \ldots, k_n) : \tilde{A}_{in}(k_1) \ldots \tilde{A}_{in}(k_n): . \tag{9}$$

Similar expansions hold for the retarded commutator

$$K^{\text{ret}}(q) = (2\pi)^{-1} \int d^4 x\, e^{iqx} \theta(x_0)\, [A(x), A(0)] \tag{10}$$

and the commutator of the "out" field

$$K^{\text{out}}(q) = (2\pi)^{-1} \int d^4 x\, e^{iqx} [A_{\text{out}}(x), A_{\text{out}}(0)] \tag{11}$$

with the coefficients $c_n^{\text{ret}}(q, k_1, ..., k_n)$ and $c_n^{\text{out}}(q, k_1, ..., k_n)$ respectively. Obviously there are relations connecting the functions f_n with the coefficients c_n^{ret}:

$$f_{n+1} = \pi(m^2 - q^2)\, (m^2 - (q-k)^2)\, c_n^{\text{ret}}(q, k_1, ..., k_n)|_{(q-k)^2 = m^2} \tag{12}$$

where

$$q = \sum_{i=1}^{n} k_i \quad \text{and} \quad k_{n+1} = q - k. \tag{13}$$

The following relations can be derived with the help of Eq. (2):
1. $c_0^{\text{out}}(q) = \varepsilon(q_0)\, \delta(m^2 - q^2)$.
2. $c_1^{\text{out}}(q, k) = 0$.
3. $c_0(q)$ is the Fourier transform of a local function for any arbitrary choice of the f_n's.

All other coefficients c have to fulfil an infinite system of nonlinear integral equations relating the coefficients c_n to the functions f_m. This set of equations is most directly derived by

(I) inserting the expansion (2) into (8),

(II) expanding the product of the two Wick monomials into the products of contractions and single Wick monomials according to Wick's theorem,

(III) comparing the result of (I) and (II) with the expansion (9).

We propose an approximation scheme for the solution of the condition deriving from locality. The approximation consists in a truncation of the expansion in Wick's theorem at a fixed number of contractions. The number of contractions taken into account defines then the order of our approximation. In the truncated equations every coefficient c_n gets only a finite number of contributions involving a finite number of f_n only. Thus, to second order the coefficients c_n are determined by only few of the functions:

c_0 by f_2,
c_1 by f_2 and f_3,
c_2 by f_2, f_3 and f_4, etc.

This approximation scheme has the virtue that it bears a close connection to intermediate states and physical thresholds: The contribution of order l to a coefficient c corresponds to intermediate states of particle number l (and a suitably chosen set of partly disconnected intermediate states of higher particle number). It is different from zero only above the threshold at $(lm)^2$.

The problem is then to find functions c_n and f_m such that the equations obtained in a certain order of approximation are fulfilled together with the Eq. (12) relating f_{n+1} and c_n^{ret}.

As we shall see in the following two sections the solutions to the first and second order approximations given there have the following properties.

1. The coefficients c_n are Fourier transforms of local functions.

2. The TCP-conditions hold (up to the same order of approximation), i.e.

$$(2\pi)^{3/2} f_n = \langle 0| j(0)| p_1, ..., p_n; \text{in} \rangle = \langle 0|j(0)| p_1, ..., p_n; \text{out} \rangle^*. \quad (14)$$

Here the operator $j(0)$ and the "out" operators, creating the "out" states, have to be understood as being defined in terms of the "in" operators as in Eqs. (2), (6) and (7).

3. $\qquad\qquad\qquad\qquad\qquad c_n^{\text{out}} = 0, \quad n \geq 1. \qquad\qquad\qquad\qquad (15)$

2. Saturation to First Order

The first order local saturation is an approximation in terms of one-particle intermediate states. The equations for the coefficients c_1 and c_2 read

$$c_1^{(1)}(q, p) = \frac{\varepsilon(q_0)\, \delta(m^2 - q^2)}{m^2 - (p - q + i\varepsilon)^2}\, f_2((p - q + i\varepsilon)^2) - \binom{q \to p - q}{p \to p}, \quad (16)$$

$$c_2^{(1)}(q, p) = \frac{\varepsilon(q_0)\, \delta(m^2 - q^2)}{m^2 - (p_1 + p_2 - q + i\varepsilon)^2}$$
$$\times f_3(\times(p_1 + p_2 + i\varepsilon)^2, (p_1 - q + i\varepsilon)^2, (p_1 + p_2 - q + i\varepsilon)^2)$$
$$+ \frac{\varepsilon(q - p_1)\, \delta(m^2 - (q - p_1)^2)}{[m^2 - (q + i\varepsilon)^2]\, [m^2 - (p_1 + p_2 - q + i\varepsilon)^2]} \times \qquad\qquad (17)$$
$$\times f_2((q + i\varepsilon)^2)\, f_2((p_1 + p_2 - q + i\varepsilon)^2)$$
$$- \binom{q \to p_1 + p_2 - q}{p_1 \to p_1,\ p_2 \to p_2}.$$

The solutions to these equations are $[q_+ = (q_0 + i\varepsilon, \boldsymbol{q})]$

$$c_1^{\text{ret}}(q, p) = \frac{\text{polynomial}}{(m^2 - q_+^2)\, [m^2 - (q_+ - p)^2]}, \qquad\qquad (18)$$

$$c_2^{\text{ret}}(q, p_1 p_2)$$
$$= \frac{\text{polynomial}}{(m^2 - q_+^2)\, [m^2 - (p_1 + p_2 - q_+)^2]} \left\{ \frac{1}{m^2 - (q_+ - p_1)^2} + (p_1 \leftrightarrow p_2) \right\}. \quad (19)$$

Thus, the solutions are uniquely determined up to polynomials. The equations for the higher coefficients c_n are of similar structure and the solutions again are essentially of pole type. The requirements (1–3) stated at the end of the last section are fulfilled as far as they apply to this order of approximation.

The first order approximation can easily be generalized to commutators of currents involving a whole set of one particle intermediate states for their saturation. Also additional constraints like exact or partial conservation of currents can easily be taken into account. The equal time structure of current commutators can be studied in particular [6].

The approximations obtained from tree graphs in the effective Lagrangian approach [7] or via the hard pion technique [8] are of the same type as the first order approximation of this section.

3. Saturation to Second Order

In the second order local approximation ($l = 2$) the question of unitarity enters the game. No general solution has as yet been given, however, the solution for $c_1(q, k)$ exhibiting locality and elastic unitarity is known.

The coefficient

$$c_1(q, p) = (2\pi)^{-1} \int d^4x \, e^{iqx} \langle 0| \, [A(x), A(0)] \, |p\rangle \tag{20}$$

is completely determined by the functions f_2 and $f_3 : (q_\pm = q \pm i\varepsilon)$

$$c_1^{(2)}(q, p) = c_1^{(1)}(q, p)$$

$$+ \left\{ \varepsilon(q_0) \, \frac{f_2(q_+^2) \, t_0^*((-q_-)^2, (p-q_-)^2)}{(m^2 - q_+^2) \, [m^2 - (p-q_-)^2]} - \binom{q \to p - q}{p \to p} \right\}. \tag{21}$$

Here $c_1^{(1)}(q, p)$, the first order approximation is given by Eq. (16); the s-wave scattering amplitude t_0^* is determined by

$$t_0^* = \tfrac{1}{2}\pi \int d^4k_1 \, [\theta(k_1^0) \, \theta(q^0 - k_1^0) + \theta(-k_1^0) \, \theta(-q^0 + k_1^0)]$$
$$\cdot \, \delta(m^2 - k_1^2) \, \delta(m^2 - (q-k_1)^2) \, f_3((-q_-)^2, (p-k_1+i\varepsilon)^2, (p-q_-)^2). \tag{22}$$

Making use of the fact that $c_1(q, p)$ is the q_0-discontinuity of $c_1^{ret}(q, p)$ and of Eq. (12) relating f_2 and c_1^{ret} we obtain the unitarity relation for f_2 as a special case of Eq. (21):

$$\text{Im} \, f_2(q_+^2) = \varepsilon(q_0) \, f_2(q_+^2) \, t(q_+^2, m^2) \quad \text{(Watson theorem)}. \tag{23}$$

The unitarity relation for t_0 can be obtained by the same manipulations from the second order approximation to $c_3(q, p_1, p_2)$. These relations turn out to be identical to the *TCP* requirements (14) for $n = 2, 3$.

The solution to the above equations is then given by (neglecting all possible polynomial dependences):

$$c_1^{\text{ret}}(q, p) = \pi^{-1} \frac{f_2(m^2) + \{f_2(q_+^2) + f_2((q_+ - p)^2) - 2 f_2(m^2)\}}{(m^2 - q_+^2)\,[m^2 - (q_+ - p)^2]} \qquad (24)$$

where $f_2(q^2)$ is determined by the expression [9]

$$f_2(q^2) = f_2(m^2) \exp\left\{ -\frac{q^2 - m^2}{\pi} \int\limits_{4m^2}^{\infty} \frac{\delta(q'^2)\,dq'^2}{(q'^2 - m^2)(q'^2 - q_+^2)} \right\}. \qquad (25)$$

Here $\delta(s)$ is the s-wave scattering phase shift as determined by t_0.

The s-wave projection of the coefficient $c_2^{\text{out}}(q, p_1, p_2)$ of the expansion of the "out" field vanishes. Thus, the four conditions stated at the end of Section 1 are fulfilled also in the approximation of second order. The solution obtained is no longer of pole type, both the functions f_2 and $c_1^{\text{ret}}(q, p)$ contain two-particle cut contributions with a threshold at $4m^2$. The function f_3 contributes only through its on shell s-wave phase $\delta(s)$, which is not restricted in this order of approximation.

A general expression for $c_1(q, p)$ and $c_1^{\text{ret}}(q, p)$ can be given in terms of an arbitrary choice of intermediate states by the following ansatz

$$c_1^{\text{ret}}(q, p) = \pi^{-1} \int dq'^2\, dQ'^2 \frac{\sigma(q'^2, Q'^2)}{(q'^2 - q_+^2)\,[Q'^2 - (p - q_+)^2]}, \qquad (26)$$

with

$$\theta(q_0')\,\theta(-Q_0')\,\sigma(q'^2, Q'^2)$$
$$= (2\pi)^{-7/2} \int d^4z\, d^4y\, e^{iq'z} e^{iQ'y} \langle 0|\, A(z)\, j(0)\, A(y)\, |0\rangle|_{(q' + Q')^2 = m^2}. \qquad (27)$$

This representation gives a local expression for $c_1^{\text{ret}}(q, p)$ for every choice of intermediate states inserted to the left and to the right of the operator $j(0)$. Even though, we know of no rigorous proof for the representation the structure of it looks very plausible: the pole residue of $c_1^{\text{ret}}(q, p)$ at $(p - q)^2 = m^2$ must reproduce $f_2(q_+^2)$ with the correct analytical properties, $c_1(q, p)$ must exhibit the $\varepsilon(q_0)$ and $\varepsilon(p_0 - q_0)$ structure and possess the reality properties contained in Eq. (21). Of course, this representation is consistent with our requirements and the approximations of first and second order, Eqs. (18) and (24), correspond to the following graphs:

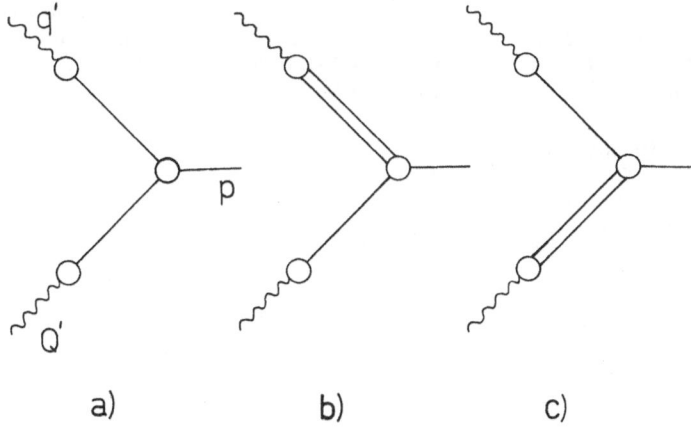

Fig. 1. Graphical representation of a first order b and c second order contributions to the spectral function

4. Conclusion

The aim of presenting a local approximation scheme in terms of intermediate states has been reached in part. The first order approximation can be carried out, the solution has the form of pole structures for the matrix elements of the retarded commutator. The second order approximation has been given for the vacuum one particle matrix element of the commutator. The problem of local approximations for matrix elements with more particles is under investigation.

References

1. Zimmermann, W.: Nuovo Cimento **13**, 503 (1959). – Völkel, A. H., Völkel, V.: Nuovo Cimento **63**A, 203 (1969), and references cited there. – Kramer, G., Meetz, K.: Commun. math. Phys. **3**, 29 (1966).
2. Symanzik, K.: J. Math. Phys. **1**, 249 (1960).
3. Tamm, I.: J. Phys. (UdSSR) **9**, 449 (1945). – Dancoff, S. M.: Phys. Rev. **78**, 382 (1950).
4. Dahmen, H. D., Rothe, K. D.: Nucl. Phys. B**15**, 387 (1970). – Stech, B.: Z. Physik **239**, 387 (1970). – Dahmen, H. D., Rothe, K. D., Stech, B.: Phys. Letters **34**B, 83 (1971). – Dahmen, H. D., Gromes, D., Rothe, K. D., Stech, B.: Phys. Letters **35**B, 335 (1971). – Stech, B.: Manuscript of a talk presented at the Symposium on Basic Questions in Elementary Particle Physics, Munich, June 8–18, 1971.
5. Haag, R.: Dan. Mat. Fys. Medd. **29**, 13 (1955). – Glaser, V., Lehmann, H., Zimmermann, W.: Nuovo Cimento **6**, 1122 (1957).

6. Dahmen, H. D., Rothe, K. D., Rothe, H. J.: Heidelberg preprint 1971, to be published.
7. Coleman, S., Wess, J., Zumino, B.: Phys. Rev. **177**, 2239 (1969) and references cited there.
8. Schnitzer, H. J., Weinberg, S.: Phys. Rev. **164**, 1828 (1967).
9. Omnès, R.: Nuovo Cimento **8**, 316 (1958).

Hans D. Dahmen
Institut für Theoretische Physik der Universität Heidelberg
D-6900 Heidelberg und
Kernforschungszentrum Karlsruhe, Germany

Duality in Deep Inelastic Electroproduction

P. V. LANDSHOFF

Contents

Introduction

The concept of duality takes various forms. It has been applied to high-energy electroproduction in three ways; in historical order, these are:

a) To build up at least part of the structure functions from an infinite sum of resonances, through a Veneziano-like amplitude.

b) To relate the behaviour of the structure functions at extremely high values of v, q^2 to what is found at more moderate values

c) In a parton model, where the partons are identified as quarks. Here duality provides a dynamical input, which gives information about the relative magnitudes of the contributions from quarks and antiquarks.

I shall talk about a) and c), leaving b) to be discussed by Dr. Rubinstein.

Although the first attempt [1] to explain the data for $v W_2^{eP}$ through a Veneziano-like model preceded the other applications of duality, at present a realistic approach by means of Veneziano-like models has to be supplemented by the information contained in the parton approach [2]. The parton model determines the relative importance of $I = \frac{1}{2}$ and $I = \frac{3}{2}$ baryon trajectories, and also of isoscalar and isovector photons; at present we do not know how to do this using the Veneziano approach by itself. Also the parton model shows us how to incorporate non-resonance, pomeron-exchange contributions.

One might be worried about identifying the partons as quarks, if quarks do not exist as physical particles. The excuse for doing so is that, when duality is incorporated, this is the only way at present known in which we obtain results that are in fair agreement with all known data on the following experiments:

  (i) deep inelastic electroproduction on a proton [3],
  (ii) deep inelastic electroproduction on a neutron [3],
  (iii) deep inelastic neutrino scattering [4],
  (iv) electron-positron annihilation [5],
  (v) muon-pair production in nucleon-nucleon scattering [6].

There is of course a theoretical problem, if quarks do not exist. However, the mass of the quarks does not appear in the results of the calculations, and when the parton model is formulated [8] through a smoothed-out field theory, it does not seem necessary for the quark propagator to have poles. That is, it is possible that the quarks appear just as fields, and not as physical particles in asymptotic states. In more phenomenological language, perhaps the quarks are bound very tightly to their parent hadrons, and cannot escape. We certainly do not understand whether this is theoretically possible, but it is not obviously impossible.

I have said that the results of the theory are in fair agreement with all known data. However, in most of the experiments the errors quoted are huge. It is worth saying that one should not expect the agreement to be perfect when the experimental errors get smaller. This is because duality is not an exact notion, being essentially dynamical; its predictions for purely hadronic reactions are in no more than fair agreement with experiment.

Quark-Parton Model

In the quark-parton model, the electromagnetic current and the weak-interaction current couple directly to quarks and antiquarks. Whether one treats the partons phenomenologically [9] or by the smoothed-out field theory [8], in the deep inelastic limit the dominant contribution to the virtual Compton amplitude arises from diagrams having the structure of Fig. 1. There the internal line is either a quark (Fig. 1a) or an antiquark (Fig. 1b) and there is a sum of six such diagrams, corresponding to the photons coupling to each type of quark p, n, λ and to each type of antiquark $\bar{p}, \bar{n}, \bar{\lambda}$. So if I write in explicitly the factors arising from the quark charges, then in an obious notation

$$F_2^{eP} = (\tfrac{2}{3})^2 (F^p + F^{\bar{p}}) + (\tfrac{1}{3})^2 (F^n + F^{\bar{n}} + F^\lambda + F^{\bar{\lambda}}) . \tag{1}$$

There is a similar expression for νW_2^{eN}. If I use charge symmetry to relate the (p, n, λ) contributions for the proton to the (n, p, λ) contributions for the neutron, this is

$$F_2^{eN} = (\tfrac{2}{3})^2 (F^n + F^{\bar{n}}) + (\tfrac{1}{3})^2 (F^p + F^{\bar{p}} + F^{\lambda} + F^{\bar{\lambda}}).\qquad(2)$$

In terms of quark fields, the weak interaction current is

$$J_\mu^{\text{weak}} = \bar{p}\,\gamma_\mu (1 - \gamma_5)\,n\qquad(3)$$

where, for simplicity, I have set the Cabibbo angle equal to zero. From this,

$$\begin{aligned}F_2^{\nu P} &= 2(F^n + F^{\bar{p}})\\ F_2^{\bar{\nu} P} &= 2(F^p + F^{\bar{n}}).\end{aligned}\qquad(4)$$

In (4) the quark labels identify the quarks attached to the bubble in Fig. 1. If all the quark propagators are equal, the identity of the upper quark that propagates between the two current vertices does not affect

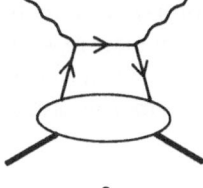
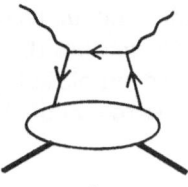

a   b

Fig. 1

the numerical value of the contribution. Thus in the first term of $F_2^{\nu P}$, an n quark is emitted by the proton, is struck by the positively-charged current and changes into a p, which is changed back into an n when it emits the second current. In the second term the sequence is $(\bar{p}, \bar{n}, \bar{p})$. The factor 2 arises because each of these two sequences contributes two equal terms, one arising from γ_μ coupling at each vertex, and the other from $\gamma_\mu \gamma_5$ at each vertex. (The cross terms, with γ_μ at one vertex and $\gamma_\mu \gamma_5$ at the other, contribute only to the structure functions F_3, and not to F_2).

Now make a dynamical assumption, based on the two-component theory of duality that is familiar in the study of hadronic reactions [10]. Each amplitude is supposed to consist of two parts:

$$R + \mathbb{P}\qquad(5)$$

where $\mathbb{P}$ corresponds to pomeron exchange, while contributions to R are very small if they correspond to duality diagrams that contain a closed loop. Here this means that, so far as their contributions to the part R is concerned, the antiquark terms and the λ-quark term in (1), (2) and (4) are negligible. These terms correspond to duality diagrams

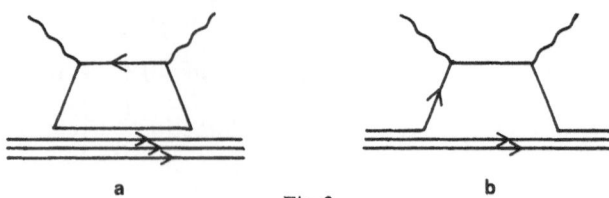

Fig. 2

of the structure of Fig. 2a, which have a closed loop; the duality diagrams (Fig. 2b) for the n and p quark terms do not. Notice that the duality diagrams are introduced just for book-keeping, and have no dynamical significance beyond this. I write

$$F_R^{\bar{n}} = F_R^{\bar{p}} = F_R^{\bar{\lambda}} = F_R^{\lambda} = 0 \,. \tag{6}$$

The duality diagrams say nothing about the diffractive part $\mathbb{P}$ of an amplitude, and in fact quarks and antiquarks contribute equally to this. This is because the pomeron has even signature, so that its contribution to the bubble in Fig. 1a does not change when we apply crossing to that bubble to obtain Fig. 1b. Since the pomeron is also isoscalar, we thus have

$$
\begin{aligned}
F_{\mathbb{P}}^{p} &= F_{\mathbb{P}}^{n} = F_{\mathbb{P}}^{\bar{p}} = F_{\mathbb{P}}^{\bar{n}} = D(\omega) \quad \text{say.} \\
F_{\mathbb{P}}^{\lambda} &= F_{\mathbb{P}}^{\bar{\lambda}} = D^{\lambda}(\omega) \quad \text{say.}
\end{aligned} \tag{7}
$$

In phenomenological language, when we calculate the part R of an amplitude we may imagine that the nucleons consist just of three quarks, as in the most naive quark model. But for the diffractive part $\mathbb{P}$ we must also take into account a sea of equal numbers of quarks and antiquarks. Notice that I am not going to make any detailed assumptions about the nature of this sea – it may be that some of the quarks and antiquarks in it are bound together to form "gluons", but this will not matter.

Since in the naive quark model the proton contains twice as many p quarks as n quarks,

$$F_R^{n} = \tfrac{1}{2} F_R^{p} = R(\omega) \quad \text{say.} \tag{8}$$

[More strictly, what can be shown is

$$\int\limits_1^\infty (\mathrm{d}\omega/2\omega)\,(F_R^{p} - F_R^{\bar{p}}) = 2, \quad \int\limits_1^\infty (\mathrm{d}\omega/2\omega)\,(F_R^{n} - F_R^{\bar{n}}) = 1\,. \tag{9}$$

These equations follow because it can be shown that applying the operation $\int\limits_1^\infty \mathrm{d}\omega/2\omega$ to the sum of Fig. 1a and Fig. 1b results in the difference

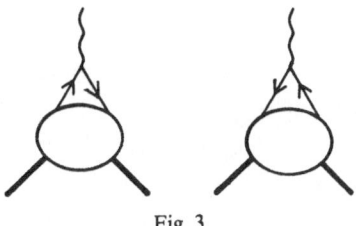

Fig. 3

between the formfactor diagrams of Fig. 3, and the value of this difference at $t = 0$ is known for each type of quark, from the isospin, charge and baryon number of the proton. We use (6) to remove the antiquark terms from (9), and then (8) follows if we make the additional assumption that the momentum distributions of the p and n quarks in the nucleons have the same shape. Verification of this last assumption comes from the scaling property of the elastic form factors of the nucleons, though this is not a proof because the momentum distribution of the quarks come into the elastic form factors in a slightly different way from that in which they contribute to Fig. 1a].

So now I have

$$F_2^{eP}(\omega) = R(\omega) + \tfrac{10}{9} D(\omega) + \tfrac{2}{9} D^\lambda(\omega)$$
$$F_2^{eN}(\omega) = \tfrac{2}{3} R(\omega) + \tfrac{10}{9} D(\omega) + \tfrac{2}{9} D^\lambda(\omega)$$
$$F_2^{vP}(\omega) = 2 R(\omega) + 4 D(\omega) \tag{10}$$
$$F_2^{\bar{v}P}(\omega) = 4 R(\omega) + 4 D(\omega).$$

I can find the function $R(\omega)$ in two ways. One is to try to calculate it; I will talk about this later. The other is to get it directly from experimental data:

$$F_2^{eP} - F_2^{eN} = \tfrac{1}{3} R(\omega). \tag{11}$$

The data [3] for the difference $(F_2^{eP} - F_2^{eN})$ is not yet very good (Fig. 4); for the present I shall pretend that it corresponds to the curve that I have labelled "$m = 3$" in the figure. Then the part $R(\omega)$ for F_2^{eP} corresponds to the curve drawn in Fig. 5. This figure also shows the data [3], and I should like to interpret the difference between the data and the curve as the diffractive contribution $(\tfrac{10}{9} D(\omega) + \tfrac{2}{9} D^\lambda(\omega))$. We know quite generally that the contribution to F_2 from the exchange of a reggeon of intercept α_0 is proportional to $\omega^{\alpha_0 - 1}$ for large ω, and so for the pomeron it should be constant for large ω. The diffractive contribution has been calculated [11] in a Veneziano-like model in which it is dual to cuts; it comes out the right shape, but its magnitude is a free parameter. If

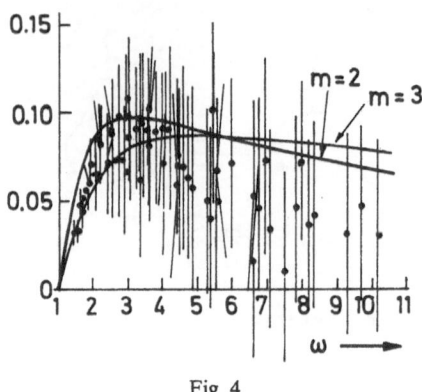

Fig. 4

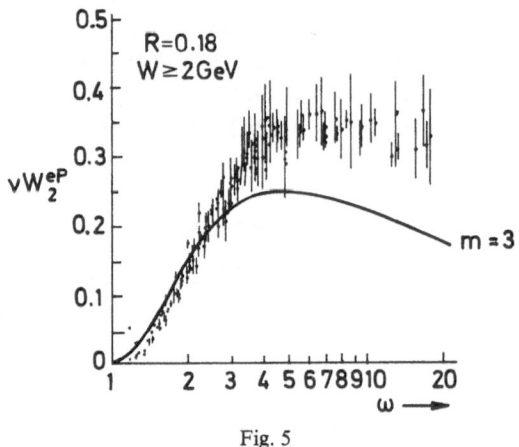

Fig. 5

the pomeron coupling obeys $SU3$ symmetry, $D^\lambda(\omega) = D(\omega)$. With this extra assumption I can now calculate $F_2^{\nu P}$ and $F_2^{\bar\nu P}$. I can also [2] calculate $F_3^{\nu P}$ and $F_3^{\bar\nu P}$ from $R(\omega)$ and $D(\omega)$. I obtain, for the total cross-section for neutrinos scattering on propane,

$$0.47 \, G^2 M E/\pi \quad \text{per nucleon} \tag{12}$$

and the experiments [4] at present give

$$(0.52 \pm 0.13) \, G^2 M E/\pi \quad \text{per nucleon.} \tag{13}$$

There is similar agreement [2] with the details of the neutrino experiment, so far as they are known. The numerical predictions are not sensitive to the assumption of $SU3$ symmetry for the pomeron coupling.

Muon-Pair Production

Consider now the experiment [6]

$$P + P \rightarrow \mu^+ \mu^- + \text{hadrons} \qquad (14)$$

where the total-energy variable s and the muon-pair mass variable $\sqrt{q^2}$ are both large, such that

$$\tau = q^2/s \qquad (15)$$

is not large. Drell and Yan [6] argued from the parton model that in the asymptotic limit the differential cross-section takes the form

$$d\sigma/d\sqrt{q^2} = (8\pi\alpha^2/3)(q^2)^{-3/2}\psi(\tau) \qquad (16)$$

and they identified one contribution to $\psi(\tau)$ as arising from the diagram in Fig. 6. Here one proton emits a quark and the other an antiquark; these annihilate to produce a virtual photon, which in turn produces

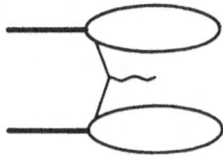

Fig. 6

the muon pair. Drell and Yan calculated the contribution to $\psi(\tau)$ from this term, and in the notation that I introduced in (1) obtain from it

$$\psi(\tau) = \sum_a Q_a^2 \int d\omega_1 d\omega_2 F^a(\omega_1) F^{\bar{a}}(\omega_2) \delta(\omega_1\omega_2 - \tau^{-1}). \qquad (17)$$

Here the summation is over all the types of quark and antiquark, and Q_a is the charge of the quark a. The limits of integration over ω_1 and ω_2 have to be chosen to correspond to the kinematical conditions of the experiment. From the foregoing analysis of electroproduction I know all six functions F^a and can plug them into (17). I obtain a different result from Drell and Yan because, as we have seen, the antiquark functions are rather smaller than the quark functions F^p and F^n, since the former contribute just to the diffractive part of electroproduction, while Drell and Yan took all six F^a to be equal. The result is the curve B drawn in Fig. 7; this figure also shows the data [6]. The curve stops at $\sqrt{q^2} = 4$ because after that it is rather sensitive to the details of how the diffractive contribution to electroproduction behaves near $\omega = 1$; all I can say for sure is that it drops very rapidly indeed between the two vertical lines.

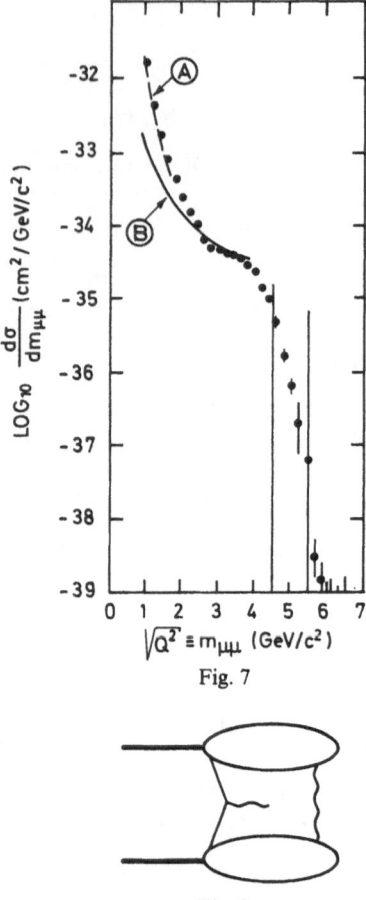

Fig. 7

Fig. 8

The analysis of the parton model in the smoothed-out field theory approach reveals [7] a contribution to $\psi(\tau)$ from a second term, Fig. 8, in which a pomeron is exchanged after the protons have emitted the partons. (That is, it is associated with "wee" partons.) Strictly, this contribution is probably not a function of τ alone, but also of log s. The details will depend on the unknown nature of the pomeron. However, as log s is a slowly-varying function, we can forget it for practical purposes. We cannot calculate the contribution from Fig. 8, because we do not know enough about the pomeron. But one can show [7] that for small τ, the contribution takes the form*

$$\tau^{-1}\tilde{\psi}(\tau) \tag{17}$$

* *Note added in proof:* This statement is now withdrawn; see Nuclear Physics B **36**, 642 (1972).

where $\tilde{\psi}(\tau)$ is regular at $\tau = 0$. Between $\sqrt{q^2} = 1$ and $\sqrt{q^2} = 2$, varies between about 0.02 and 0.08 in the experiment, so this term probably dominates here. If I choose $\tilde{\psi}(\tau)$ to be a constant to agree with the left-most experimental point, I obtain the curve A in Fig. 7. I expect the curve to continue to fall for larger values of $\sqrt{q^2}$. Remembering the logarithmic scale, we see that the sum of the two curves A and B agress well with the data.

Veneziano-like Models

In the original formulation of the two-component theory of duality, it was hoped that the part R would be constructed mainly from resonances. Although there are indications now from hadron physics that cuts are likely to be very important too [10], in electroproduction the only known way to calculate $R(\omega)$ is by means of resonances.

The general problem of constructing Veneziano-like amplitudes for currents is very hard [12]. Particular obstacles are the requirements arising from current conservation, and from factorisation of pole residues on all trajectories, that is on both the parent and on every daughter. Fortunately, there are reasons to suppose that these requirements have little connection with the properties of the amplitudes in the scaling limit, so that a much more simple-minded construction [1] suffices.

The reasons for this are, first, that W_2 arises from the double-helicity-flip virtual Compton amplitude and current conservation merely relates this amplitude to others, for which we do not need to construct models [13]. Second, the factorisation problem imposes conditions on the integrand of the representation of the amplitude in a particular part of the integration space; the scaling property arises from properties of the integrand in an altogether different part of the integration space, namely that part which gives the Fubini-Gell-Mann-Dashen fixed pole.

There is, of course, no general argument that such a fixed pole should be associated with the Compton amplitude. However, we see from (10) that, because of the dynamical assumptions that I made, the function $R(\omega)$ that arises in electroproduction also comes into the neutrino amplitudes. In fact, from (10),

$$R(\omega) = \tfrac{1}{2}\left(F_2^{\bar{v}P}(\omega) - F_2^{vP}(\omega)\right). \tag{18}$$

The amplitude $A(s, t, q_1^2, q_2^2)$ whose imaginary part gives $(W_2^{\bar{v}P} - W_2^{vP})$:

$$W_2^{\bar{v}P} - W_2^{vP} = 2M \, \mathrm{Im} \, A(s, 0, q^2, q^2) \tag{19}$$

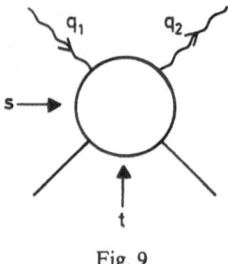

Fig. 9

(the variables are defined in Fig. 9) has $I=1$ in the t-channel, and is just the amplitude that has the fixed pole:

$$A \sim -2\,F_D(t)/\pi s$$

$$s \text{ large}, \ t, q_1^2, q_2^2 \text{ fixed} \tag{20}$$

where F_D is the Dirac isovector elastic form factor of the nucleon. Although the scaling limit is a different asymptotic limit,

$$s \to \infty, \ q_1^2 = q_2^2 = q^2, \ s/q^2 = 1 - \omega \tag{21}$$

the presence of this fixed pole, together with the analyticity properties associated with duality, plays a crucial role in ensuring that there is scaling in the scaling limit. (In the parton model also, there is a close connection between the fixed pole and scaling).

Because of crossing, I can write

$$A(s, t) = \tilde{A}(s, t) - \tilde{A}(u, t) \tag{22}$$

and identify $A(s, t)$ as the $\bar{v}P$ amplitude, or alternatively as its resonance part, since the diffractive part cancels between the two terms of (20). Then I attribute to $\tilde{A}$ the structure

$$\tilde{A}(s, t) = \sum_B V_B(s, t) + \sum_{B'} V_{B'}(u, t)$$
$$+ \sum_{B, B'} V_{BB'}(s, u). \tag{23}$$

Here B, B' denote the various baryon trajectories $N_\alpha, \varDelta$ etc. that can be put in the s, u channels of the $\bar{v}P$ amplitude. The functions $V_B(s, t)$ will be constructed to have the following properties:

a) Poles at $\alpha(q_1^2) = 1, 2, 3, \ldots$ and $\alpha(q_2^2) = 1, 2, 3, \ldots$ where α is the exchange-degenerate $(\varrho, \omega, f, A_2)$ trajectory. As the currents carry spin one, the poles at $\alpha = 2, 3, \ldots$ correspond to the spin-one daughters of the leading trajectory resonances.

b) Poles at $\alpha(t) = 2, 3, \ldots$ (There are no poles at $\alpha(t) = 0, 1$ in the double-flip amplitude).

c) Poles in the s-channel corresponding to the trajectory B

d) Regge asymptotic behaviour $t^{B(s)-\frac{1}{2}}$ when $t\to\infty$ with s, q_1^2, q_2^2 fixed

e) Regge asymptotic behaviour $s^{\alpha(t)-2}$ when $s\to\infty$ with t, q_1^2, q_2^2 fixed, *together with* a term

$$-\phi_B(t)/\pi s \tag{24}$$

$V_{B'}(u, t)$ will have similar properties. The fixed poles of all the terms in (23), together with those of $A(u, t)$, must add up so as to give the behaviour (20):

$$\sum_B \phi_B(t) - \sum_{B'} \phi_{B'}(t) = F_D(t) \tag{25}$$

(remember that $u \sim -s$ in the limit (20)).

Consider the function

$$\int_0^\infty (du_1 du_2 dz/u_1 u_2 z)(1+1/u_1)^{\alpha(q_1^2)-1}(1+1/u_2)^{\alpha(q_2^2)-1}$$
$$(1+1/z)^{\alpha(t)-2}(1+zX(u_1, u_2, z))^{B(s)-\frac{1}{2}} Y(u_1, u_2, z). \tag{26}$$

I require this integral to converge for negative values of the exponents, so when either u_1 or u_2 is large the function X is bounded and Y tends to zero. If also X and Y are neither zero nor infinite when the integration variables tend to zero or when $z\to\infty$, the integral has the same pole structure in the variables q_1^2, q_2^2, s and t as is desired for the amplitude $V_B(s, t, q_1^2, q_2^2)$. The s and t channels are dual to each other, their poles being associated with different ends of the same integration z, and their resonances produce Regge pole terms in the appropriate asymptotic limits. For example, when [14] $B(s)\to -\infty$ with t, q_1^2, q_2^2 fixed, the dominant contribution to the integral arises from the region of integration where zX is small. The Regge pole term arises from small z; to extract it make the change of variables $z = -z'/B(s)$, take the limit under the integral, so that for example

$$(1+zX(u_1, u_2, z))^{B(s)} \to e^{-z'X(u_1, u_2, 0)} \tag{27}$$

and perform the z' integration:

$$\Gamma(2-\alpha(t))(-B(s))^{\alpha(t)-2}\int_0^\infty (du_1 du_2/u_1 u_2)(1+1/u_1)^{\alpha(q_1^2)-1}$$
$$(1+1/u_2)^{\alpha(q_2^2)-1}(X(u_1, u_2, 0))^{\alpha(t)-2} Y(u_1, u_2, 0). \tag{28}$$

The functions X and Y must be chosen in such a way that the integral (26) has also the term (23) in its s-channel asymptotic behaviour. As the coefficient of s^{-1} in this term depends on t but not on q_1^2 or q_2^2, the term must arise from the part

$$u_1, u_2 \to \infty$$

$$z \text{ finite} \tag{29}$$

of the integration region. That is, the function X must tend to zero in this region. A choice of the function X that has this property is

$$X(u_1, u_2, z) = \left[1 + \frac{u_1 u_2}{e^z + u_1 + u_2}\right]^{-1}. \tag{30}$$

There will not be an attempt here to establish the essential uniqueness of this choice of X, as ultimately the choice depends on what is desired from the spectrum of daughter particles; the analysis of this is altogether beyond the scope of the present work. However, it will be remarked that the choice of X is very severely constrained by several other requirements, in addition to those already mentioned on its behaviour at the ends of the range of integration. Among these is a requirement that there be no further structure in the s or t channel asymptotic behaviour (so that, for example, X does not tend to zero when u_1 alone is large) and that the function goes to zero exponentially when $s \to \infty$ at fixed $u = (p_1 - q_2)^2$. Another natural requirement is that $X(0, 0, z) = Y(0, 0, z) = 1$, so that the amplitude for the purely hadronic process

$$\varrho^- + p \to \varrho^- + p$$

extracted from the residue of the twofold pole at $\alpha(q_1^2) = \alpha(q_2^2) = 1$ be an ordinary Veneziano beta function. In order that the residues of the s channel poles have the appropriate factorization properties, it is also required that $X(u_1, u_2, \infty)$ and $Y(u_1, u_2, \infty)$ reduce to suitably factorized forms in u_1 and u_2. The details of this are unimportant here, but it will be remarked that the form factors obtained by factorizing the s channel pole residues can be made to vanish as rapidly as one pleases for large current "mass". An explicit example was given in Ref. [1].

I still have to make the correct function $\phi_B(t)/\pi$ appear as the coeffizient of $(-s)^{-1}$ in (23). Make use of the known analyticity properties of $\phi_B(t)$ to write it as the transform:

$$\phi_B(t) = \int_0^\infty (dz/z)(1 + 1/z)^{\alpha(t)-1} \tilde{\phi}_B(z). \tag{31}$$

(By making the change of variable $1 + 1/z = e^x$ one can see that this is essentially a Laplace transform). Choose the function Y such that in the region (29)

$$Y \sim (\alpha'/\pi)(1 + z)(u_1 + u_2)^{-1} \tilde{\phi}_B(z). \tag{32}$$

Then on making the changes of variable

$$u_1 = -B(s)\, u_1'$$
$$u_z = -B(s)\, u_2', \tag{33}$$

taking the limit under the integral and performing the u_1', u_2' integrations, we find the result (23).

Consider now the Bjorken scaling limit (21). In this limit the factors in (26) that involve q_1^2 and q_2^2 reduce to finite exponentials when one makes the changes of variable (33) and takes the limit under the integral [14]; that is the dominant contribution again arises from the region of integration (29). Using (32), we can perform the u_1' and u_2' integrations and find the asymptotic form

$$-\phi_B(t, \omega)/s \tag{34}$$

where

$$\phi_B(t, \omega) = \pi^{-1} \int_0^\infty dz\, \frac{(1 + 1/z)^{\alpha(t) - 1}\, \tilde{\phi}_B(z)}{z + (1 - \omega)^{-1}}. \tag{35}$$

If $\phi_B(t)$ is real for $\alpha(t) < 1$, its transform $\tilde{\phi}_B(z)$ is real. Hence $\phi_B(0, \omega)$ is real for $\omega < 1$. If it is now continued to $\omega > 1$ it acquires an imaginary part due to the vanishing of the denominator in the integrand:

$$\mathrm{Im}\, \phi_B(0, \omega) = -\theta(\omega - 1)\, \omega^{\alpha(0) - 1}\, \tilde{\phi}_B((\omega - 1)^{-1}). \tag{36}$$

The existence of this imaginary part implies that $\phi_B(0, \omega)$ has a branch point at $\omega = 1$, even though (26) has been constructed to have only poles, and no branch points. The explanation of this is familiar: the apparent cut is synthesised by the infinite numbers of poles in the variables s and q^2. The absence of these poles in fact prevents the asymptotic behaviour (34) from being strictly valid in a wedge containing the positive ω real axis. But it is supposed that it would be valid on the real axis in a unitarised theory, where the trajectory functions have cuts and the poles are displaced from the real axis.

Because of (25), when I add together the contributions from all the terms in $A(s, t)$ I obtain in the Bjorken limit (21)

$$A(s, t, q^2, q^2) \sim -\phi(t, \omega)/s + \phi(t, -\omega)/u \tag{37}$$

where $\phi(t, \omega)$ is given by an integral like (35), but with $\tilde{\phi}_B(z)$ replaced by $\tilde{F}_D(z)$ defined by

$$F_D(t) = \int_0^\infty (dz/z)\, (1 + 1/z)^{\alpha(t) - 1}\, \tilde{F}_D(z). \tag{38}$$

I have assumed that the terms $V_{BB'}(s, u)$ do not contribute in this limit; certainly it is hard to construct a Veneziano-like model in which they do. Because of (36) the first term of (37) has an imaginary part in the region $\omega > 1$ of interest, with the final result

$$R(\omega) = \omega^{\alpha(0)} (\omega - 1)^{-1} \tilde{F}_D ((\omega - 1)^{-1}). \tag{39}$$

The curves in Fig. 4 and 5 correspond to the choice

$$F_D(t) = \frac{\Gamma(1 - \alpha(t))}{\Gamma(1 + m - \alpha(t))} \cdot \frac{\Gamma(1 + m - \alpha(0))}{\Gamma(1 - \alpha(0))} \tag{40}$$

which, for large t, has the behaviour $F_D(t) \sim t^{-m}$. This gives

$$R(\omega) = \frac{\Gamma(1 + m - \alpha(0))}{\Gamma(1 - \alpha(0))\,\Gamma(m)}\, \omega^{\alpha(0) - m} (\omega - 1)^{m - 1}. \tag{41}$$

References

1. Landshoff, P. V., Polkinghorne, J. C.: Nuclear Physics B **19**, 432 (1970). Some of the assumptions of this paper were relaxed in later work; in particular see Reference [2].
2. — — Nuclear Physics B **28**, 240 (1971); Phys. Letters **34** B, 621 (1971).
3. Bloom, E. D., *et al.*: Kiev Conference report (1970) SLAC-PUB-796.
4. Myatt, G., Perkins, D. H.: Phys. Letters **34** B, 542 (1971).
5. Bartoli, B., *et al.*: Nuovo Cimento 70 A, 615 (1970). For the theory see, N. Cabibbo, G. Parisi and M. Testa, Nuovo Cimento Letters **4**, 35 (1970).
6. Christenson, J., *et al.*: Phys. Rev. Lett. **25**, 1523 (1970). For the theory, see S. D. Drell and T. M. Yan, Phys. Rev. Letters **25**, 1523 (1970) and Ref. [7].
7. Landshoff, P. V., Polkinghorne, J. C.: Nuclear Physics B **33**, 221 (1971).
8. — — Short, R. D.: Nucl. Physics B **28**, 225 (1971).
9. Bjorken, J. D., Paschos, E. A.: Phys. Rev. **185**, 1975 (1969); Llewellyn Smith, C. H.: Nucl. Physics B **17**, 277 (1970); West, G. B.: Phys. Rev. Letters **23**, 1415 (1969).
10. Harari, H.: Phys. Rev. Letters **22**, 1078 (1969); **26**, 1400 (1971).
11. Atkinson, D., Contogouris, A. P.: Bonn preprint.
12. For a review, see Ademollo, M.: Springer Tracts Mod. Physics **59**, 135 (1971).
13. That is, except at $q^2 = 0$, where there is a constraint on the double-flip amplitude alone; but $q^2 = 0$ is far from the scaling limit.
14. Throughout, asymptotic limits are taken in directions in the complex plane for which the integrals converge. It is implicitly assumed that no extra undesirable terms appear when asymptotic limits of the analytic continuations are taken.

Professor Dr. P. V. Landshoff
Department of Applied Mathematics and
Theoretical Physics
University of Cambridge
Cambridge CB 3 9 EW/England
Silverstreet

Parton Models of Inelastic Lepton Scattering*

C. H. Llewellyn Smith

Contents

1. Introduction and Outline

The large number of speakers who will deal with different aspects of highly inelastic lepton scattering at this summer school is evidence for the importance currently attributed to this subject. As the first, I will begin on an elementary level by discussing the kinematics in some detail and making a few preliminary remarks which should provide some background for the following lectures for those who are not familiar with the subject.

Next I will turn to the parton model. I will outline the usual "derivation" of the model but the philosophy adopted subsequently will be that the parton model provides a useful heuristic device from which results can be abstracted which might be true more generally. In fact I

---

* Work supported by the U.S. Atomic Energy Commission.

will then show that these results can be rederived formally using infinite momentum commutators (although superficially different this part of my lectures will overlap with those of Gross and Jackiw). However, it should be noted that this formal derivation of the parton model is not necessarily superior to the traditional one – the difficulties are simply buried in formal (possibly meaningless) expressions.

In the last section I give references to papers about processes which I shall not discuss and make a remark about the use of local duality in inelastic lepton scattering.

2. Preliminary Remarks

a) Electroproduction

Kinematics

We shall consider the process in which an electron scatters from a nucleon and only the final electron is observed. To lowest order in α, this is represented by the Feynman diagram:

where k, k', q, p and p_F are four momenta, E and E' are lab. energies and θ is the lab. scattering angle.

With states normalized covariantly,

$$\langle p | p' \rangle = 2 E (2\pi)^3 \delta^3(\boldsymbol{p} - \boldsymbol{p'})$$

the differential cross section is given by

$$d\sigma = (2M2E)^{-1}|M|^2(2\pi)^{-3}(2E')^{-1}d^3k'(2\pi)^4\delta^4(k+p-k'-p_F) \quad (1)$$

according to the usual rules.

The invariant matrix element is

$$M = e^2 \bar{u}(k', s') \gamma_\mu u(k, s) (i/q^2) \langle p_F | J_\mu(0) | p \rangle$$

where $\bar{u}u = 2m_e$ and we use the metric $q^2 = q_0^2 - \boldsymbol{q}^2$. On squaring and averaging (summing) over initial (final) spins we obtain:

$$|M|^2(2\pi)^4\delta^4(k+p-k'-p_F) = (4\pi e^4/q^4)\, m_{\mu\nu} W^{\mu\nu} \quad (2)$$

where $q = k - k'$,

$$
\begin{aligned}
m_{\mu\nu} &= \overline{\sum_{s}} \sum_{s'} \bar{u}(k,s)\,\gamma_\mu u(k',s')\,\bar{u}(k',s')\,\gamma_\nu u(k,s) \\
&= \tfrac{1}{2} \mathrm{Tr}\,(\not{k}+m_e)\,\gamma_\mu(\not{k'}+m_e)\,\gamma_\nu \\
&= 2(k_\mu k'_\nu + k_\nu k'_\mu - g_{\mu\nu} k\cdot k') + O(m_e^2)
\end{aligned}
\tag{3}
$$

$$
W_{\mu\nu}(F) = \tfrac{1}{2}\overline{\sum}\,\langle p|\,J_\mu^+(0)\,|F\rangle\,\langle F|\,J_\nu(0)\,|p\rangle\,(2\pi)^3\,\delta^4(q+p-p_F)
$$

and $\overline{\sum}$ denotes a spin average over the initial states, which will be understood henceforth.

If we sum over all the final states F, we may write:

$$
\begin{aligned}
W_{\mu\nu} &= \sum_F W_{\mu\nu}(F) \\
&= -(q_{\mu\nu} - q_\mu q_\nu/q^2)\,W_1 + (p_\mu - q_\mu q\cdot p/q^2)(p_\nu - q_\nu q\cdot p/q^2)\,W_2
\end{aligned}
\tag{4}
$$

$$
W_i = W_i(\nu, q^2), \quad \nu = q\cdot p = M(E-E').
$$

This is the most general tensor which we can construct out of the available vectors which respects current conservation $(q^\nu W_{\mu\nu} = W_{\mu\nu} q^\nu = 0)$ and parity conservation. Combining Eqs. (1)–(4) and putting $m_e = 0$ we obtain:

$$
\begin{aligned}
d^2\sigma/d|q^2|\,d\nu &= (\pi/M E E')\,d^2\sigma/d\Omega\,dE' \\
&= (4\pi\alpha^2/q^4 M^2)(E'/E)(W_2\cos^2\theta/2 + 2W_1\sin^2\theta/2).
\end{aligned}
\tag{5}
$$

We shall collect some useful properties of the W_i^γ which describe electroproduction below when we consider the structure functions which describe neutrino production.

Scale Invariance

We have defined $W_{\mu\nu}$ (Eqs. (3) and (4)) so that it is dimensionless. Consider a theory in which all masses and coupling constants with dimensions are zero, so that no natural scale is defined and the theory is invariant under scale transformations $x \to \lambda x$ and $p \to p/\lambda$. If we consider the form of $W_{\mu\nu}$, we see that in such a theory we would have to make the replacements:

$$
\begin{aligned}
W_1 &\to F_1(\omega) \\
W_2/M^2 &\to F_2(\omega)/\nu \\
(\omega &= 1/x = 2\nu/(-q^2))
\end{aligned}
\tag{6}
$$

in Eq. (4).

Bjorken proposed [1] that Eq. (6) might hold in the real world in the limit $\nu \to \infty$ with ω fixed, in which case masses might be irrelevant.

In fact it seems that it holds to quite a good approximation at rather modest values of v and q^2. The relevant experimental results are discussed by Drees [2]. Here we only remark that in perturbation theory

$$v W_2/M^2 \xrightarrow[\substack{v \to \infty \\ \omega\text{fixed}}]{} F_2(\omega) + f(\omega) \log(q^2/M^2). \tag{7}$$

The second term is not excluded by the SLAC data, although it turns out that $f/F \lesssim 1/10$, if we fit the SLAC-MIT data [3] in the range $1 \text{ Gev}^2 < |q^2| < 10 \text{ Gev}^2$ with this form for $\omega = 4$.

Henceforth we assume that Bjorken scaling holds, i.e. that $f(\omega) \equiv 0$. Note that

a) If this is not the case, the rest of these lectures are empty.

b) If it is the case, arguments based on perturbation theory may be irrelevant in the scaling region.

b) Neutrino Production

The differential cross section for neutrino production can be calculated as above with

$$e^2/q^2 \to G/\sqrt{2}$$

$$W_{\mu\nu}^{\nu,\bar{\nu}} = \tfrac{1}{2} \sum \sum_F \langle p| J_\mu^{\mp}(0) |F\rangle \langle F| J_\nu^{\pm}(0) |p\rangle (2\pi)^3 \delta^4(q+p-p_F) \tag{8}$$

$$m_{\mu\nu}^{\nu,\bar{\nu}} = \text{Tr}\,k\gamma_\mu(1 \mp \gamma_5)(k'+m_\mu)\gamma_\nu(1 \mp \gamma_5).$$

Quite generally $W_{\mu\nu}$ may be written:

$$W_{\mu\nu}^{\nu,\bar{\nu}} = -g_{\mu\nu} W_1^{\nu,\bar{\nu}} + M^{-2} p_\mu p_\nu W_2^{\nu,\bar{\nu}} - (2M^2)^{-1} i \varepsilon_{\mu\nu\alpha\beta} p^\alpha q^\beta W_3^{\nu,\bar{\nu}}$$
$$+ M^{-2} q_\mu q_\nu W_4^{\nu,\bar{\nu}} + (2M^2)^{-1}(q_\mu p_\nu + q_\nu p_\mu) W_5^{\nu,\bar{\nu}} \tag{9}$$
$$+ (2M^2)^{-1} i(q_\mu p_\nu - q_\nu p_\mu) W_6^{\nu,\bar{\nu}}.$$

Whatever the form of the lepton current, $m_{\mu\nu}$ may be expanded in a similar way with $p_\mu \to k_\mu$. Hence $m_{\mu\nu} W^{\mu\nu}$ is a second order polynomial in $k \cdot p = M E_\nu^{\text{lab}}$ and:

$$E_\nu^2(d^2\sigma/dq^2 dv) \sim m_{\mu\nu} W^{\mu\nu} \sim A(v, q^2) + B(v, q^2) E_\nu + C(v, q^2) E_\nu^2. \tag{10}$$

This result [4] (sometimes referred to as a consequence of "locality") depends only on the assumption of a first order vector interaction:

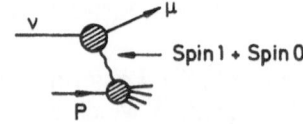

It therefore allows a test of this assumption. From our point of view it is important because it implies that only three combinations of structure functions can be obtained without polarization measurements.

In the conventional case $W_{1,2,3,}$ are effectively determined, since the other structure functions are multiplied by q_μ and $q_\mu l^\mu \sim O(m_\mu)$. Explicitly, Eqs. (8) and (10) give:

$$d^2\sigma^{\nu,\bar{\nu}}/d|q^2|\,d\nu = (G^2/2M\pi)(E'/E)[W_2^{\nu,\bar{\nu}}\cos^2\theta/2 + 2W_1^{\nu,\bar{\nu}}\sin^2\theta/2 \quad (11)$$
$$+ W_3 M^{-1}(E+E')\sin^2\theta/2] + O(m_\mu^2)W_{4,5}.$$

In principle $W_{4,5,6}$ can be determined by measuring the muon's polarization (the explicit expressions are given, e.g., in Ref. [5]) but this is hard, since it is necessarily 100% longitudinally polarized in the approximation $m_\mu = 0$ in the $V - A$ theory.

For some purposes it is convenient to define the total cross sections for the fictitious process of current − nucleon scattering. First we define the normalized, conserved polarization vectors:

$$\varepsilon_\mu^S = (-q^2)^{-1/2}(q_3, 0, 0, q_0)$$
$$\varepsilon_\mu^R = 2^{-1/2}(0, 1, i, 0)$$
$$\varepsilon_\mu^L = 2^{-1/2}(0, i, 1, 0) \quad (12)$$
$$[q = (q_0, 0, 0, q_3)].$$

Contracting $\varepsilon_\mu^* \varepsilon_\nu$ with $W^{\mu\nu}$ and putting in the appropriate factors we obtain:

$$\sigma_{R,L} = \frac{\pi}{\nu + q^2/2}(W_1 \pm \tfrac{1}{2}\sqrt{\nu^2/M^4 - q^2/M^2}\,W_3)$$
$$\sigma_S = \frac{\pi}{\nu + q^2/2}(W_2(1 - \nu^2/M^2 q^2) - W_1) \quad (13)$$

where, according to an arbitrary convention, we have used the flux factor for a zero mass ($q^2 = 0$) current with the same $s = (p+q)^2$.

According to Regge theory, or our knowledge of real cross sections, we might expect that

$$\sigma \xrightarrow[q^2\text{fixed}]{\nu\to\infty} \nu^{\alpha-1},$$

where $\alpha(t)$ is the leading Regge trajectory which can be exchanged in elastic current-nucleon scattering. Hence

$$W_1 \xrightarrow[q^2\text{fixed}]{\nu\to\infty} \nu^{\alpha_1}$$
$$W_2 \to \nu^{\alpha_2-2} \quad (14)$$
$$W_3 \to \nu^{\alpha_3-1}.$$

Presumably $\alpha_1 = \alpha_2 = 1$ since the Pomeron can be exchanged. However, since W_3 is the $V - A$ interference term (it multiplies the only pseudotensor in Eq. (9)) it requires odd G parity exchange in the t channel, if only first class currents are present. Therefore we expect $\alpha_3 \approx 1/2$.

Since $\sigma_{R,L,S} \geq 0$, according to their definition, we see that:

$$W_2(1 - v^2/M^2 q^2) \geq W_1 \geq (2 M^2)^{-1} \sqrt{v^2 - M^2 q^2} \, |W_3| \geq 0. \qquad (15)$$

In fact, the complete exploitation of the fact that $W_{\mu\nu}$ is a positive semi-definite Hermitean matrix is somewhat more complicated. The necessary and sufficient condition that the diagonal elements remain positive semidefinite under all unitary transformations is that all the subdeterminants of $W_{\mu\nu}$ be positive semidefinite. This yields two more nonlinear inequalities involving W_4, W_5 and W_6 [6].

Some other useful properties which follow immediately from the definition of $W_{\mu\nu}$ are:

1. $W_{\mu\nu} = W_{\nu\mu}^* \rightarrow W_i$ real
2. If T is conserved, $W_{\mu\nu}(p_\alpha, q_\beta) = W^{\mu\nu}(p^\alpha, q^\beta)$

$$W_{1\ldots 5} - \text{real}$$

$$W_6 - \text{imaginary} \therefore W_6 \equiv 0.$$

(The only effect of W_6 is to give the final muon a polarization out of the reaction plane. It is easy to see directly that this transverse polarization is forbidden by T invariance in the approximation $\alpha = e^2/4\pi = 0$ — see, e.g., Ref. [5].)

3. The $W_i^{\nu, \bar{\nu}}$ have no kinematical singularities or zeros. However, since $W_{\mu\nu}$ is finite as $q^2 \rightarrow 0$ the W_i^γ satisfy

$$W_2^\gamma \xrightarrow{q^2 \to 0} O(q^2)$$

$$(W_1^\gamma + v^2 W_2^\gamma/M^2 q^2) \sim O(q^2). \qquad (16)$$

4. We can define different structure functions by summing separately over strange and non-strange final states and put

$$W_i^{\nu, \bar{\nu}} \rightarrow W_i^{\nu, \bar{\nu}} \cos^2 \theta_C + \omega_i^{\nu, \bar{\nu}} \sin^2 \theta_C$$

$$F_i^{\nu, \bar{\nu}} \rightarrow F_i^{\nu, \bar{\nu}} \cos^2 \theta_C + f_i^{\nu, \bar{\nu}} \sin^2 \theta_C \qquad (17)$$

where θ_C is the Cabbibo angle. The assumption of the charge symmetry condition for the $\Delta s = 0$ weak current:

$$J_\lambda^+ = - e^{-i\pi I_2} J_\lambda^- e^{i\pi I_2}$$

gives:

$$\langle F(I, I_3)| J_\lambda^\pm |p\rangle = \langle F(I, -I_3)| J_\lambda^\mp |n\rangle.$$

On summing over all states F the isospin reflection is irrelevant and we obtain

$$W_i^{vp} = W_i^{\bar{v}n}, \quad W_i^{\bar{v}p} = W_i^{vn}$$
$$F_i^{vp} = F_i^{\bar{v}n}, \quad F_i^{\bar{v}p} = F_i^{vn}. \tag{18}$$

5.

$$W_{\mu\nu}^{v;\bar{v}} = (4\pi)^{-1} \int d^4x \langle p| J_\mu^{\mp}(x) J_\nu^{\pm}(0) |p\rangle e^{iq\cdot x} \quad \text{(a)}$$
$$= (4\pi)^{-1} \int d^4x \langle p| [J_\mu^{\mp}(x), J_\nu^{\pm}(0)] |p\rangle e^{iq\cdot x} \quad \text{(b).} \tag{19}$$

It is easy to check that the additional term is the second line makes no contribution in the physical region $v > 0$. Eq. (19b) defines the W_i for $v < 0$ according to the crossing relations:

$$W_i^v(v, q^2) = -W_i^{\bar{v}}(-v, q^2)$$
$$W_i^v(v, q^2) = -W_i^v(-v, q^2), \tag{20}$$

which are derived by examining $W_{\mu\nu}^*$.

The advantage of this definition of $W_{\mu\nu}$ for $v < 0$ is that the optical theorem holds for all v:

$$W_{\mu\nu} = (2\pi)^{-1} \operatorname{Im} T_{\mu\nu}$$
$$T_{\mu\nu}^{v,\bar{v}} = i \int d^4x \, \theta(x_0) \langle p| [J_\mu^{\mp}(x), J_\nu^{\pm}(0)] |p\rangle e^{iq\cdot x} + \text{Polynomial in } v. \tag{21}$$

$T_{\mu\nu}$ is, of course, the amplitude for forward current scattering so that Eq. (21) is the usual relation between forward scattering and the total cross section:

$$\operatorname{Im} \; \rangle\!\!\subset\!\!\langle \; \sim \; \sum_F \left| \rangle\!\!\subset\!\!\!\in F \right|^2$$

We may write

$$T_{\mu\nu} = -g_{\mu\nu} T_1 + M^{-2} p_\mu p_\nu T_2 + \cdots \tag{22}$$

in correspondence with Eq. (9). The $T_i(v, q^2)$ are analytic functions in the cut v plane which satisfy the dispersion relations:

$$T_{1,4}(v, q^2) = T_{1,4}(0, q^2) + 2v \int_{-\infty}^{+\infty} \frac{W_{1,4}(v', q^2) \, dv'}{v'(v' - v)}$$
$$T_{2,3,5}(v, q^2) = 2 \int_{-\infty}^{+\infty} \frac{W_{2,3,5}(v', q^2)}{v' - v} \, dv', \tag{23}$$

assuming Regge asymptotic behaviour (Eq. (14)). (The equations for $T_{1,4}$ appear superficially divergent. However, using the crossing properties (Eqs. (20)) and the fact that $W^v - W^{\bar{v}}$ does not receive contributions from the Pomeron, we see that in fact they are convergent.)

Scale Invariance

In neutrino production, Bjorken scaling obviously takes the form

$$W_1 \xrightarrow[\omega\,\text{fixed}]{\nu\to\infty} F_1(\omega)$$

$$\nu W_i/M^2 \to F_i(\omega) \qquad (i \neq 1).$$

Present data do not allow these predictions to be tested directly. However, there are two direct consequences of scaling, which can be confronted with the data already. The first is the behaviour of the total neutrino cross section. In a theory without any parameters with dimension the only possibility is $(s = (p + q)^2)$

$$\sigma \sim G^2 s$$

since $G^2 \sim M^{-4}$. The second is for the average Q^2 at fixed s which must satisfy:

$$\langle Q^2 \rangle \sim s.$$

Both these predictions are compatible with the data [6, 7], although the values of s involved are rather small.

3. Parton Model

a) Intuitive Derivation

The parton model consists essentially of applying the impulse approximation to inelastic lepton scattering, the partons being Feynman's name for the constituents of the nucleon [8, 9]. The impulse approximation is supposed to apply when

$$\Delta E_{\text{interaction}} \gg \Delta E_{\text{binding}}$$

or, in coordinate space,

$$\tau_{\text{interaction}} \ll \tau_L$$

where τ_L is the lifetime of the virtual states of the nucleon. In this case it is argued that the constituents may be treated as free particles during the interaction. The necessary condition is achieved by considering a frame in which the proton's momentum is very large ($|\boldsymbol{p}| \to \infty$) so that τ_L is lengthened by time dilation while $\tau_{\text{int}} \sim 1/q_0 \ll \tau_L$.

A further condition is that the scattering from the constituents is incoherent. This is achieved by assuming that the partons momenta transverse to the nucleon's momentum is sharply cut off (which is not unreasonable since the debris produced when the nucleon is broken up by collision with another hadron satisfies this condition) and considering conditions in which $|q_\perp|$ is very large compared to the cut off. In this case diagrams in which the current is absorbed by different partons cannot interfere (final state interactions being ignored since they turn one set of complete states into another – as we shall see in detail below).

To see that these conditions may be satisfied in Bjorken's limit, and to put the discussion above on a more quantitative basis, we consider briefly the arguments of Drell, Levy and Yan [10], who considered a field theory of pions and nucleons endowed ab initio with a cut off in transverse momentum (we refer to their papers for complete details).

They use the interaction representation in which the complete Heisenberg current operator $J_\mu(x)$ is related to the free field current operator $j_\mu(x)$ by:

$$J_\mu(x) = U^{-1}(t) j_\mu(x) U(t)$$

where

$$U(t) = T \exp - i \int_{-\infty}^{t} H_I(t') \, dt'.$$

Therefore:

$$W_{\mu\nu} = \tfrac{1}{2} \overline{\sum_F} \sum \langle Up| j_\mu(0) U(0) |F\rangle \langle F| U^{-1}(0) j_\nu(0) |Up\rangle (2\pi)^3 \qquad (24)$$
$$\cdot \delta^4(q+p-p_F)$$

$$|Up\rangle = U(0)|p\rangle$$

$$= \sqrt{Z_2} \left\{ |p\rangle + \sum' \frac{|n\rangle \langle n| H_I |p\rangle}{E_p - E_n} \right.$$
$$\left. + \sum'_{n,m} \frac{|m\rangle \langle m| H_I |n\rangle \langle n| H_I |p\rangle}{(E_p - E_m)(E_p - E_n)} + \cdots \right\}$$

where Σ' indicates summation over all states except $|p\rangle$ and we recall that in this "old fashioned perturbation theory" the virtual particles are on mass shell and momentum is conserved at each vertex but not energy.

If $|q_\perp| \to \infty$, then in any finite order of perturbation theory the only diagrams which contribute have the form:

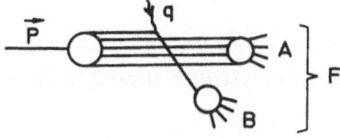

since the transverse momentum cut off requires an infinite number of exchanges to connect the groups A and B. We may therefore write $|F\rangle = |A\rangle |B\rangle$ and $U|F\rangle = U|A\rangle U|B\rangle$.

Consider now a state $|n\rangle$ in $U|p\rangle$ which has n constituents with momenta:

$$p_i = \eta_i p + k_i$$
$$k_i \cdot p = 0, \quad \sum_i \eta_i = 1.$$

Then

$$\lim_{|\boldsymbol{p}|\to\infty} E_p - E_n = (1/2|\boldsymbol{p}|)\left(M^2 - \sum_i (k_i^2 + M_i^2)/\eta_i\right) + O(1/\boldsymbol{p}^2), \quad \text{it} \quad 0 < \eta_i < 1$$
$$= O(|\boldsymbol{p}|) \quad \text{otherwise,} \tag{25}$$

provided $|k_i|$ is cut off so that this expansion makes sense. The energy denominators in Eq. (24) therefore favour the intermediate states with $0 < \eta_i < 1$ by a factor $\boldsymbol{p}^2$ provided this is not compensated by factors of $|\boldsymbol{p}|$ from the numerator. It is possible to show [10] that such a compensation does not occur in the diagrams which contribute to $W_{\mu\nu}$ provided we restrict our attention to j_0 and the component of $\boldsymbol{j}$ which is parallel to $\boldsymbol{p}$.

Similarly if we consider $U|F\rangle = U|A\rangle U|B\rangle$ when $|q_\perp| \to \infty$, we find that

$$E_{UF} - E_F \sim O(1/|\boldsymbol{p}|) + 0(1/|\eta\,\boldsymbol{p} + \boldsymbol{q}|)$$

in the leading diagrams.

Hence, as $|\boldsymbol{p}| \to \infty$ we may replace $q_0 + E_p - E_F$ by $q_0 + E_{Up} - E_{UF}$ in the δ-function in Eq. (24). In this case, in frames in which $|q_\perp| \to \infty$ and $|\boldsymbol{p}| \to \infty$:

$$\lim_{|\boldsymbol{p}|\to\infty} W_{\mu\nu} = (4\pi)^{-1} \int d^4x \langle Up|j_\mu(x)j_\nu(0)|Up\rangle\, e^{iq\cdot x}$$

and we have "derived" the parton model: the bare current scatters incoherently (because of the transverse momentum cut off) from free on-mass-shell constituents (which have positive momenta along $\boldsymbol{p}$) and energy and momentum are conserved at the elementary vertex.

Drell and Yan [10] have analysed the set of frames in which this picture holds. The traditional choice is the ep centre of mass in which:

$$q_0 = (2\nu + q^2)/4|\boldsymbol{p}|$$
$$|q_\perp| = \sqrt{-q^2}$$

$|\boldsymbol{p}|$ and $\nu \to \infty$ with $\omega = 2\nu/(-q^2)$ and q_0 fixed, but this choice is without deep significance, since the electron is merely a device to provide a virtual photon.

b) Parton Model Calculations

According to the discussion above the structure functions are given by adding the contribution of each type of constituent in the parton model and

$$W_k(\nu, q^2) = \int \sum_i U_i(x)\, W_k^i(x, \nu, q^2)\, dx \tag{26}$$

where $U_i(x)$ is the probability of finding a type i parton with a fraction x of the proton's momentum p and W_k^i is its contribution to W_k represented by the diagram

We have put $p_\mu^i \simeq x p_\mu$ since $|p| \to \infty$ and $|k_i|$ is cut off. The free particle kinematics yield the on-mass-shell δ-function $\delta(2xq \cdot p + q^2)$ which gives scaling.

More explicitly, for a free particle

$$\langle p_i' | J_\mu^\gamma(0) | p_i \rangle = Q_i(p + p')_\mu + \text{spin dependent terms}$$

$$\sum_F |F\rangle \langle F| = (2\pi)^{-3} \int |p_F\rangle \, d^4 p_F \, \delta(p_F^2 - m_F^2) \langle p_F|$$

$$W_{\mu\nu}^\gamma = \tfrac{1}{2} Q_i^2 (2p + q)_\mu (2p + q)_\nu \delta(q^2 + 2q \cdot p) + \text{spin dependent terms.}$$

The spin dependent terms vanish as $q \to 0$, so that putting $p_\mu \to x p_\mu$ we find:

$$W_2^i = (1/x) [2x^2 Q_i^2 \delta(q^2 + 2xq \cdot p)]$$

where $1/x$ is put in to compensate for the different normalizations per unit volume of the proton ($\sim 2E$) and the parton ($\sim 2E_i = 2xE$) states.

Hence

$$\nu W_2^\gamma = \sum_i x u_i(x) Q_i^2$$

$$(x = 1/\omega = -q^2/2\nu).$$

(27)

In order to proceed we must decide:
1. What are the partons?
2. What are the $u_i(x)$?

Experiment already provides a clue about the spin of the partons. Consider the parton photon interaction in the Breit frame

If the parton has spin zero it carries no angular momentum in or out along q. Hence it cannot absorb a transverse photon and ($\sigma_T = \sigma_R + \sigma_L$):

$$R = \sigma_S/\sigma_T = \infty \quad \text{(spin 0).}$$

If the parton has spin $\tfrac{1}{2}$ its helicity is unchanged by the electromagnetic interaction in the limit $|q| \to \infty$ and therefore, since its direction is reversed, it must absorb one unit of spin from the photon. Hence $\sigma_S = 0$ in this case and

$$R = \sigma_S / \sigma_T = 0$$

$$\text{or} \quad 2F_1 = \omega F_2 \quad (\text{spin } \tfrac{1}{2}). \tag{28}$$

Experimentally it is known [3] that R is small and can be fitted by $R = -q^2/v^2$ which is zero in Bjorken's limit. Therefore the majority, and perhaps all, of the charged partons must have spin $\tfrac{1}{2}$.

Since we wish to consider models in which Gell-Mann's current algebra hypthesis holds, it is rather natural to assume that the charged partons are quarks. We consider the results of this model in the next section. First we collect some results which follow from assuming the conventional weak quark current

$$J_\mu^+ = \bar{p} \gamma_\mu (1 - \gamma_5)(n \cos\theta_C + \lambda \sin\theta_C).$$

For $\Delta S = 0$ processes on free quarks this implies that:

$$W_{\mu\nu}^{\nu n, \nu \bar{p}} = \tfrac{1}{4} \operatorname{Tr}(\not{p} + M) \gamma_\mu (1 \mp \gamma_5)(\not{p} + \not{q} + M) \gamma_\nu (1 \mp \gamma_5) \delta((p+q)^2 - M^2).$$

For large v this gives

$$W_1^{\nu n, \nu \bar{p}} = \delta(x + 2v/q^2)$$

$$v W_2^{\nu n, \nu \bar{p}}/M^2 = 2x\delta(x + 2v/q^2) \tag{29}$$

$$v W_3^{\nu n, \nu \bar{p}}/M^2 = \mp 2\delta(x + 2v/q^2)$$

for the W_k^i is Eq. (26).

4. The Quark Parton Model

a) A Difficulty

An immediate difficulty with the quark parton model is that, according to the discussion above, the final state should consist of two separate groups of particles (A and B) each of which has non-integral baryon number and charge. Since quarks are (presumably) not being produced at SLAC, we must avoid this impasse or reject the model as irrelevant at present energies. A possible escape route is provided by the "wee partons" which carry a vanishing fraction of the proton's momentum as $|p| \to \infty$. These "wees", which we ignored above, have little sense of direction and therefore one which originates in group A may be assimilated by group B, and vice – versa. Drell and Yan [10] argue that, although the wees play no role in finite order perturbation theory, they may be important if all orders are summed. Their suggestion derives some support from the work of Chang and Yan [11] who found that in a ϕ^3 theory (albeit in a different situation) the "wees" contributed

$$\frac{1}{s} \sum_n \frac{G^n \ln s^n}{n!} = s^{G-1} > 0 \quad \text{as} \quad s \to \infty \quad \text{if} \quad G > 1.$$

For the moment we take refuge in a different philosophy – namely that we are using the quark parton model as the simplest model available in which scaling and current algebra obtain, from which we can abstract results which might be true more generally even if the naive discussion here is invalid. However, Dashen has advanced arguments which suggest that this difficulty persists in the formal approach, which may imply the existence of real quarks, but not necessarily that they are being produced at SLAC; we refer to Ref. [12] for a discussion of this point.

b) General Results of the Quark Parton Model

In this section we will subscribe to the philosophy outlined immediately above and see what results can be obtained in the quark parton model, if we keep the distributions $U_i(x)$, which characterize the model as general as possible. The results will be rederived formally in Section 5.

We already found (Eq. (28)) that

$$2F_1 = \omega F_2,$$

since quarks have spin $\frac{1}{2}$. Another immediate result is that, since scale invariance and chiral invariance are both broken by the quark mass, chiral invariance will obtain in the scaling region. The tensor $W_{\mu\nu}$ will therefore be conserved, in which case:

$$F_4^{\nu,\bar{\nu}} = 0 \qquad 2F_1^{\nu,\bar{\nu}} = F_5^{\nu,\bar{\nu}}. \tag{30}$$

(The first result follows from the positivity conditions for $W_{\mu\nu}$ together with Eq. (28). It is interesting to observe that these conditions give $\nu W_6 = 0$ if either Eq. (28) or Eq. (30) obtains.)

We shall only consider the $\Delta s = 0$ structure functions here (the result for the $\Delta s = 1$ structure functions are given in Ref. [5]). Using Eqs. (28), (29) and (18) we see that we are left with 6 structure functions, which we may take to be:

$$F_1^{\gamma p, \gamma n, \nu p, \nu n}, \qquad F_3^{\nu p, \nu n},$$

which can be calculated in terms of the six distributions for quarks and antiquarks in the proton: $u_p, u_n, u_\lambda, u_{\bar{p}}, u_{\bar{n}}$ and $u_{\bar{\lambda}}$ (the distributions in the neutron are directly related by an isospin rotation). The λ quarks do not contribute to $\Delta s = 0$ weak interactions and in electromagnetic interactions they clearly contribute in the combination $u_\lambda + u_{\bar{\lambda}}$. Effectively, therefore, we have six structure functions determined by five distributions. We can evidently obtain one relation which is [13][1]:

$$12(F_1^{\gamma p} - F_1^{\gamma n}) = F_3^{\nu p} - F_3^{\nu n}.$$

$$\text{(Sakata, Fermi, Yang: } 4(F_1^{\gamma p} - F_1^{\gamma n}) = F_3^{\nu p} - F_3^{\nu n})$$

(31)

as is easily checked using Eqs. (26)–(29). Here and below we give the result of the Sakata and Fermi Yang models to indicate the model dependence of our results.

We already noted that $u_\lambda + u_{\bar\lambda}$ contributes to electromagnetic scattering only. Since $u_i \geqq 0$, being a probability, we can therefore obtain the useful inequality [13]:

$$F_1^{\nu p} + F_1^{\nu n} \leqq 18(F_1^{\gamma p} + F_1^{\gamma n})/5$$

$$\text{(Sakata-Fermi Yang: } F_1^{\nu p} + F_1^{\nu n} = 2(F_1^{\gamma p} + F_1^{\gamma n})).$$

(32)

In the Sakata (Fermi-Yang) model the λ is neutral (does not exist) and there is therefore an equality since the six structure functions are determined by 4 distributions.

This exhausts the local relations. We now exploit the fact that the u_q must yield the correct values for the proton's conserved quantum numbers so that:

$$\text{Strangeness} = 0 = \int dx (u_\lambda - u_{\bar\lambda}),$$

$$\text{Charge} = \int dx [\tfrac{2}{3}(u_p - u_{\bar p}) - \tfrac{1}{3}(u_n - u_{\bar n})],$$

$$\text{Baryon number} = \tfrac{1}{3} \int dx (u_p + u_n - u_{\bar p} - u_{\bar n}).$$

The first relation is useless in $\Delta S = 0$ reactions, since $u_\lambda - u_{\bar\lambda}$ is not measured. Using Eqs. (29) and (27), the second relation gives:

$$\int (F_2^{\nu n} - F_2^{\nu p}) \, dx/x = 2.$$

(33)

This is the scaling limit of the Adler sum rule [14], which depends only on the isovector nature of the weak current (and the validity of the derivation) but not on the nature of the constituents. Using the spin $\tfrac{1}{2}$ relation $2F_1 = \omega F_2$ it may be rewritten:

$$\int (F_1^{\nu n} - F_1^{\nu p}) \, dx = 1.$$

(34)

This is the Bjorken "backward" sum rule, which depends on the spin of the constituents. (The right hand side is zero in the spin 0 parton model and in the algebra of fields).

---

[1] In this paper the lowest moments $\int x \, dx$ of Eqs. (31) and (32) were also derived in the gluon model. At that time I was not brave enough to consider $[\partial^n J/\partial t^n, J]$ which gives the moments $\int x^n dx$ for all n.

Using Eqs. (26) and (29) the baryon number condition gives

$$\int (F_3^{\gamma p} + F_3^{\gamma n})\,dx = -6$$

(Sakata – Fermi Yang: $6 \rightarrow 2$) \hspace{2cm} (35)

– a sum rule discovered by David Gross and myself [16]. The right hand side depends essentially on the non-integral baryon number attributed to the quarks; if it is correct it would verify the quarks algebra.

So far we have not exploited momentum conservation which reads:

$$1 = \int x\,dx \sum_i u_i(x) \tag{36}$$

according to the parton assumption that the constituents all move in the same direction in the $|p| \rightarrow \infty$ frame. The difficulty in using Eq. (36) is that the right hand side contains contributions from neutral gluons which may be present (and serve to bind the quarks together), whose distributions cannot be measured, since they do not contribute to lepton scattering. However, we may separate their contribution and write:

$$1 - \varepsilon = \int x\,dx \sum_{q=1}^{6} u_q(x)$$

$$0 \le \varepsilon \le 1 \tag{37}$$

where ε is the fraction of the proton's momentum carried by the gluons.

Eq. (37) may be used to obtain the interesting upper bound:

$$\int_0^1 (F_2^{\gamma p} + F_2^{\gamma n})\,dx \le \tfrac{5}{9} \tag{38}$$

which is satisfied by the data (unless big surprises await us at unexplored x) the left hand side being ~ 0.32.

A question which arises is whether ε is zero. A possible test involving electromagnetic data alone is provided by the relation [5, 17]

$$\int F_2^{\gamma p, \gamma n}\,dx \ge (1 - \varepsilon)/9 \tag{39}$$

– the left hand is certainly $> \tfrac{1}{9}$ for the proton and very likely also for the neutron (in the model of Bjorken and Paschos [9] the integrals are $\ge \tfrac{2}{9}$ if no gluons are present, but this requires more specific assumptions about the $u_q(x)$).

Finally we note that we can derive the relation [4, 17, 18]

$$\varepsilon = 1 + \int dx (3 F_2^{\gamma p + \nu n}/4 - 9 F_2^{\gamma p + \gamma n}/2). \tag{40}$$

The result of the CERN experiment gives [6, 7]

$$\sigma^{\nu p + \nu n}/2 = G^2 M E (0.52 \pm 0.13)/\pi$$

and hence

$$\int F_2^{\gamma p + \nu n} dx \geqq 1.08 \pm 0.27 . \tag{41}$$

Inserting this in Eq. (40) and using [3]

$$\int F_2^{\gamma p + \gamma n} dx = 0.28 \pm 0.04 \tag{42}$$

we find

$$\varepsilon \geqq 0.52 \pm 0.38 , \tag{43}$$

i.e. the present data seem to require gluons.

Eqs. (41) and (42) are compatible with (32). In fact the model is quite compatible with all present data [5] as is illustrated by the more specific model discussed by Landshoff in his lectures at this summer school.

A comparison of Eqs. (41) and (42) with Eq. (32) shows that the Sakata and Fermi-Yang models are excluded by the data. In fact any model with integrally charged constituents will tend to give too small a value for $F_2^{\gamma}/F_2^{\gamma}$. F_2^{γ} depends on Q_i^2 while F_2^{γ} depends on the "weak charges" which are zero or one, depending on the isospin but not on Q_i. Therefore we expect that in most models with integral charges $F_2^{\gamma}/F_2^{\gamma}$ will be less than in the quark model. Since the data (Eqs. (41) and (42)) is already approximately equal to the maximum value permitted by the quark model (Eq. (32)) we see that models with integral charges are likely to be excluded. This has been shown explicitly by Nachtmann in several cases [30].

Nachtmann has also derived several new inequalities in the quark parton model [30]. The most interesting is

$$\frac{1}{4} \leqq \frac{F_2^{\gamma n}(x)}{F_2^{\gamma p}(x)} = \frac{\frac{4}{9}(u_n + u_{\bar{n}}) + \frac{1}{9}(u_p + u_{\bar{p}} + u_\lambda + u_{\bar{\lambda}})}{\frac{4}{9}(u_p + u_{\bar{p}}) + \frac{1}{9}(u_n + u_{\bar{n}} + u_\lambda + u_{\bar{\lambda}})} \leqq 4$$

which follows from the fact that $u_i \geqq 0$. The latest data [31] give a value of the order of $\frac{1}{4}$ for this ratio (albeit with large errors) for $x \simeq 1$. More accurate measurements of this quantity will be very interesting. Nachtmann has derived further inequalities which depend on the fact that the nucleon belongs to an isodoublet or that it belongs to an octet (assuming exact $SU(3)$ symmetry). All these results can be rederived using the formal methods outlined in the next section or (equivalently) by means of light-cone commutators.

5. Formal Derivations

The formal derivation which we shall present starts from the Bjorken-Johnson-Low (BJL) expansion [19, 20] of $T_{\mu\nu}$ (Eq. (21)). In the limit

$q_0 \to i\infty$, the exponential provides a very rapidly convergent factor $\exp - |q_0| \, t$ in the t-integration. It might, therefore, be sensible to expand the commutator in a Taylor series about $t = 0$ which gives

$$T_{\mu\nu} \xrightarrow{q_0 \to i\infty} \sum_{n=0}^{\infty} (i/q_0)^{n+1} \int d\mathbf{x} \, e^{-i\mathbf{q}\cdot\mathbf{x}} \langle p| \, [\partial^n J_\mu^+(\mathbf{x}, 0)/\partial t^n, J_\nu(0)] \, |p\rangle \\ + \text{polynomial in } q_0 \tag{44}$$

provided the commutators exist. Eq. (44) is the BJL expansion. It is at this point that we part company with perturbation theory in which [21]

1. The equal time commutators do not all exist.

2. Even when they do exist, the values obtained using Eq. (44) do not coincide with the values given by naive canonical commutators.

However, as explained above, we assume that scaling occurs and this gives us courage to proceed blindly using Eq. (44) and naively manipulating operators according to the canonical rules.

In order to avoid inessential complications we consider the case of scattering a scalar current in which

$$T = i \int \theta(x_0) \, e^{i q \cdot x} \langle p| \, [J(x), J(0)] \, |p\rangle \, d^4 x + \text{polynomial in } q_0$$

$$= T(0, q^2) + 2v \int_{-\infty}^{\infty} \frac{W(v', q^2) \, dv'}{v'(v' - v)} + \cdots,$$

$$T(\omega, q^2) = T(0, q^2) + 2\omega \int_{1}^{\infty} \frac{\omega' W^+ + \omega W^-}{\omega'(\omega'^2 - \omega^2)} \, d\omega' + \cdots \tag{45}$$

$$= T(0, q^2) + 2 \sum_{n} \int \left(\frac{\omega^{2n+1} W^+}{\omega'^{2n+2}} + \frac{\omega^{2n+2} W^-}{\omega'^{2n+3}} \right) d\omega' + \cdots$$

$$W^\pm = W(\omega, q^2) \pm W(-\omega, q^2), \quad |\omega| < 1.$$

Next we can take the limit $q_0 \to i\infty$ in a frame with $\mathbf{q} = 0$ and compare the coefficients of $(1/q_0)^n$ in Eqs. (45) and (44) (assuming scaling), following which we divide by $(p_0)^n$ and let $p_0 \to \infty$. More directly; we can let $q_0 \to i\infty$ with $\omega = 2 p_0 /(-q_0)$ fixed ($|\omega| < 1$) and compare coefficients of ω which gives:

$$2 \int F^\pm d\omega/\omega^{n+2} = \lim_{p_0 \to \infty} i(-i/2 p_0)^{n-1} \int d\mathbf{x} \langle p| \, [\partial^n J(\mathbf{x}, 0)/\partial t^n, J(0)] \, |p\rangle \tag{46}$$

$$- \text{if } n \text{ odd}, \ + \text{if } n \text{ even}, \ F(\omega) = \lim_{\substack{\gamma \to \infty \\ \omega \text{ fixed}}} W(\omega, q^2).$$

In the case of vector currents the moments of different structure functions are given in essentially the same way by the commutators of different components of the currents e.g.

$$2 \int F_1^\pm \, d\omega/\omega^{n+2} = \lim_{p_0 \to \infty} i(i/2p_0)^{n+1} \int d\boldsymbol{x} \langle p_2| \, [\partial^n J_x^+(\boldsymbol{x}, 0)/\partial t^n, J_x^-(0)] \, |p_2\rangle$$
$$- \text{ if } n \text{ even}, \ + \text{ if } n \text{ odd}, \ F_1^\pm = F_1^\text{v} \pm F_2^\text{v}. \tag{47}$$

The complete set of these relations is given in Refs. [4, 17].

In order to calculate time derivatives of the currents we must specify the interaction Hamiltonian, which we take to be the sum of the renormalizable interactions:

$$H_I = g_s \overline{\psi} \psi + g_p \phi \overline{\psi} \gamma_5 \psi + g_v B_\mu \overline{\psi} \gamma^\mu \psi. \tag{48}$$

Formally we can construct the commutators

$$\int d\boldsymbol{x} [\partial^n J(\boldsymbol{x}, 0)/\partial t^n, J(0)] \tag{49}$$

completely using H_I. However, we need only pick out those pieces whose diagonal matrix can grow like p_0^{n+1} (or faster) when $p_0 \to \infty$, since only these pieces contribute in Eq. (47) and the analogous equations for the other F_i. The pieces must obviously be parts of Lorentz tensors of rank $n+1$ or higher.

For simplicity we consider first the case $g_v = 0$. We note that the equal time commutators never introduce inverse powers of masses or fields. Therefore, in Eq. (49) we have an operator of dimension M^{n+3} constructed out of positive powers of masses and fields, from which we must pick pieces of tensors of rank $\geq n+1$. Noting that the operator must contain at least two quark fields, it is easy to see that the only such operators are [5, 17]:

$$\overline{\psi}(0) \gamma_{\alpha_1} \partial_{\alpha_2} \dots \partial_{\alpha_{n+1}} \psi(0)$$
$$\overline{\psi}(0) [\gamma_{\alpha_1}, \gamma_{\alpha_2}] \partial_{\alpha_3} \dots \partial_{\alpha_{n+2}} \psi(0) \tag{50}$$

Therefore we may set $g_s = g_p = M_s = M_p = M_Q = 0$ in calculating the operators. Hence we have derived the parton model in this case, since we have shown that all moments of the structure functions (and hence the structure functions themselves, barring pathologies) are given by the matrix elements of the same one-body operators as in a free field theory of massless quarks.

We have treated the case $g_v \neq 0$ by this method elsewhere [17]. Again it turns out that we can put $M_v = 0$. Because of the gauge invariance it is then hardly surprising that we need only change the free field theory results by putting $i\partial_\mu \to i\partial_\mu + g_v B_\mu$ everywhere, which does not change any of the parton model results which depend only on the $SU(3)$ and tensor properties of the operators. We do not give details here since the vector case will be elegantly treated by David Gross in his lectures at this school.

The discussion above reestablishes all the results of Section 4 except the sum rules and inequalities based on momentum conservation. To derive them we look at the explicit form of Eq. (47) in the case $n = 1$:

$$\int F_2^\gamma dx = \lim_{p_0 \to \infty} (1/2 p_0^2) \langle p_z | \overline{\psi}(0) (i \gamma_z \partial_z + g_v \gamma_z B_z) Q^2 \psi(0) | p_z \rangle \qquad (51)$$

$$\int F_z^{vp+vn} dx = \lim_{p_0 \to \infty} (1/2 p_0^2) \langle p_z | \overline{\psi}(0) (i \gamma_z \partial_z + g_v \gamma_z B_z) (4B + 2Y) \psi(0) | p_z \rangle$$

where Q, B and Y are the usual $SU(3) \times SU(3)$ matrices and

$$Q^2 = 2B/3 + Y/6 + I_3/3 .$$

An important point in the following is that the matrices $B, B - Y$ and $2B + Y \pm 2 I_3$ make positive semidefinite contributions whenever they appear on the right hand side of Eq. (51) [13] (this is needed in the formal derivation of Eq. (32)). If we call one of these matrices λ, then we can consider a function F_2^λ defined in terms of $\overline{\psi} \gamma_\mu \lambda \psi$ just as F_2^γ is defined in terms of $\overline{\psi} \gamma_\mu Q \psi$. In the analogue of Eq. (51) for F_2^λ, Q^2 will be replaced by $\lambda^2 \alpha \lambda$. The left hand side is positive semidefinite and therefore the right hand side must be so also (in parton language this corresponds to the assumption that $u_i \geq 0$).

Next we note that

$$\lim_{p_0 \to \infty} (1/2 p_0^2) \langle p_z | \overline{\psi}(0) (i \gamma_z \partial_z + g_v \gamma_z B_z) \psi(0) | p_z \rangle$$
$$= \lim_{p_0 \to \infty} (1/2 p_0^2) \langle p_z | \theta_{zz} - \theta_{zz}^g - \mathcal{L} | p_z \rangle$$
$$= 1 - \varepsilon \geq 0$$
$$\varepsilon = \lim_{p_0 \to \infty} \langle p_z | \theta_{zz}^g | p_z \rangle / 2 p_0^2$$

where $\theta_{\mu\nu} = \delta \mathcal{L} / \delta g_{\mu\nu}$ is the symmetric energy momentum tensor and $\theta_{\mu\nu}^g$ is the part of $\theta_{\mu\nu}$ due to the free gluon fields. Clearly $\varepsilon \leq 1$, since the left hand side is positive semidefinite according to the argument above. If we could establish that $\varepsilon \geq 0$ this would give all the results at end of Section 4. Naively this seems to be the case since θ_{zz}^g is a sum of squares. However, this is not a sufficient condition since the subtraction of the vacuum expectation value is understood here (and throughout) [23].

For (pseudo-)scalar gluons we can consider scattering the current $\phi \partial_\mu \phi$ and show that their contribution to ε is positive semidefinite using an argument similar to that in the paragraph following Eq. (51). For vector gluons we have not yet been able to establish positive semidefiniteness, although the parton analogue (that all constituents move the same way in the infinite momentum frame) seems to be true in perturbation theory in a cut-off field theory [24].

To summarize: The "momentum sum rules" at the end of Section 4 can be derived in models with scalar and pseudoscalar gluons. It is possible that they can also be derived with vector gluons.

6. Miscellaneous Remarks

a) Other Processes

We refer to Refs. [10, 25–27] for a discussion of the application of parton models to some other processes.

b) Parton Models and Local Duality

It has been suggested [28, 29] that many of the properties of the structure functions near $\omega = 1$ can be calculated in terms of the behaviour of the Born term as $Q^2 \to \infty$. I am sometimes asked how this could be understood in parton models, since they deal with incoherent scattering, while the Born term involves the coherent process of elastic scattering. The answer is that it cannot be understood and in fact the parton model (or the formal calculations using $[\dot{J}, J]$ commutators) is incompatible with these calculations, as can be seen by considering spin $\frac{1}{2}$ parton models.

Consider deep inelastic $e\pi$-scattering. The arguments which are appartly successful [29] in the ep-case (see the accompanying lectures by Rubinstein) can be applied giving:

1. $$R = \left(\frac{\sigma_L}{\sigma_T} \right)_{\omega \approx 1} \sim \left(\begin{matrix} \text{Born term} \\ Q^2 \to \infty \end{matrix} \right) = \infty$$

since $\sigma_T \equiv 0$ for elastic scattering off a spin 0 particle.

2. $$S = \left(\frac{\nu W_2^{e\pi^0}}{\nu W_2^{e\pi^+}} \right)_{\omega \approx 1} \sim \left(\begin{matrix} \text{Born term} \\ Q^2 \to \infty \end{matrix} \right) = 0 .$$

In contrast, $R = 0$ in spin $\frac{1}{2}$ models and S is presumably not zero (it is one in the quark model and in all models in which there are no local isotensor operators with "twist" – equals dimension minus spin – less than or equal to two). Therefore the extreme use of local duality in inelastic lepton scattering is incompatible with constituent models or formal calculations involving the commutator $[\dot{J}, J]$.

Acknowledgements. I wish to thank B. Renner, B. Schrempp-Otto, F. Schrempp and all the other organizers of the summer school for their warm hospitality during my stay in Hamburg.

References and Footnotes

1. Bjorken, J. D.: Phys. Rev. **179**, 1547 (1969).
2. Drees, J.: Springer Tracts Mod. Phys. **60**, 107 (1971).
3. Bloom, E., et al.: SLAC-PUB 796 (presented to the Kiev converence) and SLAC-PUBS 815 and 907 (to be published in Phys. Rev.).

4. Lee, T. D., Yang, C. N.: Phys. Rev. **126**, 2239 (1962). – Pais, A.: Phys. Rev. Letters **9**, 117 (1962).
5. Llewellyn Smith, C. H.: Neutrino Reactions at Accelerator Energies, SLAC-PUB 958 (to be published in Physics Reports). This paper contains some discussion of other aspects of inelastic ν reactions (e.g. the composition of the final state) and of the result of making some (weak) dynamical assumptions in the quark parton model.
6. Doncel, M. G., De Rafael, E.: Nuovo Cimento **4** A, 363 (1971).
7. Budagov, I., et al.: Phys. Letters **30** B, 364 (1969). – Myatt, G., Perkins, D.: Phys. Letters **34** B, 542 (1971).
8. Feynman, R.: unpublished and Phys. Rev. Letters **23**, 1415 (1965) and in "High Energy Collisions". London: Gordon and Breach 1969.
9. Bjorken, J. D., Paschos, E. A.: Phys. Rev. **158**, 1975 (1969).
10. See Drell, S. D., Yan, T. M.: Ann. Phys. **66**, 555 (1971) and refs. therein.
11. Chang, S. J., Yan, T. M.: Phys. Rev. Letters **25**, 1586 (1970).
12. Fritzsch, H., Gell-Mann, M.: In Center for Theoretical Studies University of Miami Tracts in Mathematics and Natural Science Vol. 2, Gordon and Breach 1971 and in Proc. Tel-Aviv Conference on Duality and Symmetry in Hadron Physics, Weizmann Science Press of Israel 1971.
13. Llewellyn Smith, C. H.: Nucl. Phys. B **17**, 277 (1970).
14. Adler, S. L.: Phys. Rev. **143**, 1144 (1966).
15. Bjorken, J. D.: Phys. Rev. **163**, 1767 (1967).
16. Gross, D. J., Smith, C. H. Llewellyn: Nucl. Phys. B **14**, 337 (1969).
17. Llewellyn Smith, C. H.: Phys. Rev. D **4**, 2392 (1971).
18. With $\varepsilon = 0$ this is the sum rule derived independently by Fritzsch and Gell-Mann in free quark models (Ref. [12]).
19. Bjorken, J. D.: Phys. Rev. **148**, 1467 (1966).
20. Johnson, K., Low, F. E.: Prog. Theor. Phys. Supp. 37–38, 74 (1966).
21. See, e.g., Gross, D. J., Jackiw, R., Treiman, S. B.: Current Algebras and Their Applications, Princeton Univ. Press (in press).
22. Cornwall, J. M., Norton, R. E.: Phys. Rev. **177**, 2584 (1969).
23. I am indebted to Sidney Coleman for a useful discussion about this point.
24. Drell, S. D.: private communication.
25. — Yan, T. M.: Phys. Rev. Letters **25**, 316 (1970).
26. Jaffe, R. L.: SLAC PUB 913.
27. Bjorken, J. D., Paschos, E. A.: Phys. Rev. D **1**, 1450 (1970).
28. Bloom, E. D., Gilman, F. J.: Phys. Rev. Letters **25**, 1140 (1970).
29. Gilman, F. J.: In Proc. Tel-Aviv Conference on Duality and Symmetry in Hadron Physics, Weizmann Science Press of Israel 1971.
30. Nachtmann, O.: Orsay preprint LPTHE 71/29.
31. Kendall, H.: Rapporteur's talk at the Cornell Conference, Aug. 1971.

Dr. C. H. Llewellyn Smith
Stanford Linear Acceleration Centre
Stanford University
Stanford, California 94305, USA

Duality for Real and Virtual Photons*

Hector R. Rubinstein

Contents

Introduction

The purpose of these lectures is to review and discuss some known and new results concerning duality and scaling. Particular emphasis is given to questions concerning the possibility that electroproduction and photoproduction are more closely connected than expected, or more precisely, that the limit $q^2 \to 0$ is very smooth.

Most of the considerations described here are contained in papers written in collaboration with Rittenberg.

In the first part we review shortly the kinematics and notation. We discuss a few general concepts, like the relations between Regge and Bjorken limits.

In the second part we introduce the concept of duality and discuss it in the context of fixed q^2 sum rules as done originally by Bloom and Gilman.

We emphasize the difference between strong and variable q^2 processes. In the latter, duality has a meaning irrespective of the model one uses for the asymptotic form of the amplitude.

In the third part we discuss analyticity and *local* duality in the whole $v - q^2$ plane. We arrive to these ideas, or some of them by different roads (Regge, light cone and finite energy sum rules).

In the fourth section we discuss the extent to which duality is locally obeyed and introducing an appropriate scaling variable we find a remarkable connection with photoproduction. Properties of the neutron

---
* This work has been partially supported by the National Bureau of Standards, USA.

proton structure functions difference are analyzed. Questions concerning the time like region are briefly mentioned, and predictions that will test the model in depth are discussed.

We conclude by discussing the relevance of these results under the light of current available models.

Kinematics and Limiting Properties of the Structure Functions

The double differential cross-section for inelastic electron scattering for an inclusive process in which only the electron coordinates are observed is given by [1]:

$$\frac{d^2\sigma}{d\Omega' dE'} = \frac{4\alpha^2 E'^2}{q^4} [2 W_1(v_1 q^2) \sin^2 \theta/2 + W_2(v, q^2) \cos^2 \theta/2] \quad (1)$$

where E' and $d\Omega'$ caracterize the energy and solid angle of the outgoing electron, θ is the angle measured with respect to the beam direction, q^2 is the mass of the photon, that in these experiments is kept spacelike.

$$v = E - E'. \quad (2)$$

E being the initial electron energy.

The aforementioned quantities are related by the equation

$$q^2 = 4 E E' \sin^2 \theta/2. \quad (3)$$

We also compute the missing mass to be

$$s = W^2 = 2 m v + m^2 - q^2, \quad (4)$$

$$u = -2 m v + m^2 - q^2, \quad (5)$$

where m is the proton mass.

The main difference with hadronic reactions is that the external mass can be varied. The $v - q^2$ plane can then be studied by properly defining lines as seen in Fig. 1. Constant W^2 lines have positive universal slope and negative intercept, while fixed angle lines are given by the equation

$$q^2 = 4 E^2 - 4 E v \sin^2 \theta/2. \quad (6)$$

We also introduced constant ω_W lines defined by

$$q^2 = \frac{2 m v + m^2}{\omega_W} - a^2 \quad (7)$$

where a^2 is a constant.

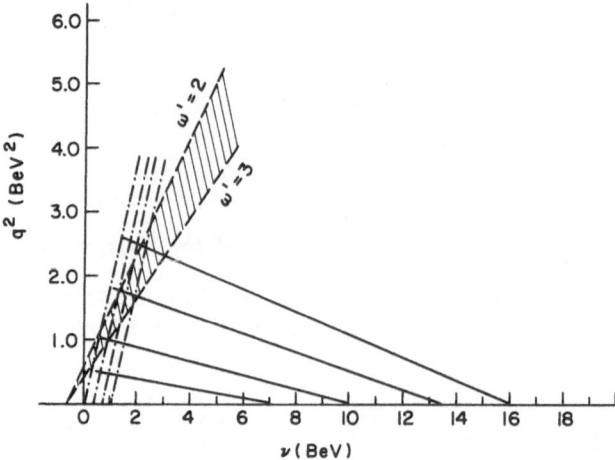

Fig. 1. The $v - q^2$ plane. The quasi-horizontal lines correspond to $\theta = 6°$, $E = 7$, 10, 13, 5 and 16 GeV discussed in Ref. [2]. Dotted lines correspond to fixed ω', the other to fixed s-values

The structure functions W_1 and W_2 are the absorptive parts of the forward Compton scattering amplitude (spin averaged). The formulae reads

$$W_1 = [(v - q^2/2m)/4\pi^2\alpha] \, \sigma_T, \tag{8}$$

$$W_2 = [(v - q^2/2m)/4\pi^2\alpha] \, \frac{q^2}{q^2 + v^2} (\sigma_T + \sigma_S) \tag{9}$$

where σ_S is the scalar and σ_T the transverse photon cross-section. Notice that W_2 is kinematically constrained to vanish at $q^2 = 0$. This simply reflects the transversality of physical photons. Since σ_T is finite for $q^2 = 0$ we have

$$\lim q^2 \to 0 \; \frac{W_2}{q^2} = \frac{1}{v} \, \frac{\sigma_T(q^2 = 0)}{4\pi^2\alpha} \neq 0 \tag{10}$$

an expression we will use later.

Another quantity of importance is

$$R = \frac{\sigma_S}{\sigma_T} = \frac{v + q^2/2m}{W_1} \left[\frac{q^2 W_2}{q^2 + v^2} \right]. \tag{11}$$

The Regge and Bjorken limits are different asymptotic domains of the structure functions. It has been conjectured that they may be related [2] though there is at present little or no evidence that this is the case.

We first keep q^2 fixed and consider

$$\lim v \to \infty \, W_1 \sim \beta_1(0, q^2) \, v^{\alpha(0)-1} + O(v^{\alpha(0)-2}), \qquad (12)$$

$$\lim v \to \infty \, W_2 \sim \beta_2(0, q^2) \, v^{\alpha(0)-2} + O(v^{\alpha(0)-3}), \qquad (13)$$

is the $t = 0$ intercept of the leading trajectory that couples to forward compton scattering like the Pomeron or A_2, for example. β's are residue functions and we have not put, notice, a Regge scale factor. The exponents are different because W_1 and W_2 have different number of helicity flips.

Since these functions are absorptive parts, contact terms and fixed poles make feel their presence only indirectly and through dispersion relations [3].

The Bjorken limit obtains when

$$q^2 \text{ and } v \to \infty \quad \frac{2mv}{q^2} = \omega = \omega_W = \text{finite} . \qquad (14)$$

Scaling is the statement that these limit exist

$$W_1 \xrightarrow[v_1 q^2 \to \infty]{} F_1(\omega), \qquad (15)$$

$$v \, W_2 \xrightarrow[v_1 q^2 \to \infty]{} F_2(\omega). \qquad (16)$$

In perturbation theory this conjecture is not verified [4]. One obtains scaling up to logarithms. This problem is to be kept in mind. A stronger connection has been proposed by some authors [2]. The same term given by (12) and (13) that is leading in the Regge limit is leading in the Bjorken limit. This implies

$$\beta_1(q^2) \sim (q^2)^{-\alpha(0)-1}, \qquad (17)$$

$$\beta_2(q^2) \sim (q^2)^{-\alpha(0)-2}. \qquad (18)$$

This behaviour obtains in simple ladder theories [5] (neglecting logarithms!) though counter examples exist as well. We will not assume these connections in our analysis but it will be helpful to use the language here and there. Using light cone analysis it is also clear that the Bjorken and Regge limits obtain support from different regions in configuration space [6]. Not surprisingly, the Bjorken limit tests short distances while Regge distances are related to the masses of the hadrons involved, or sometimes to the universal slope $1 \, (\text{GeV})^2$.

Fixed q^2 Duality

We will assume in what follows that Bjorken scaling holds for the proton data to a reasonable degree of accuracy (see Fig. 2). Hence, we assume

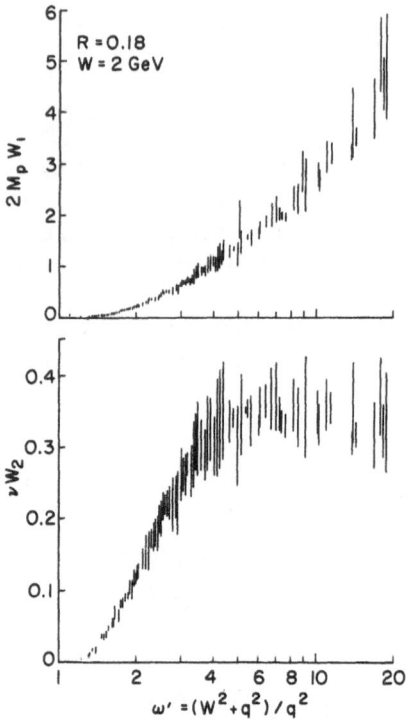

Fig. 2. Values of W_1 and νW_2 plotted against ω' for $W = 7, 2$ GeV and varying q^2 for $R = 0.18$

(12) and (13) to be valid. For large q^2 and ν, in the neutron data this is satisfied provided some additional assumptions are made [7].

Duality in electroproduction carries slightly different connotations as compared to purely hadronic processes. In the latter case one makes a statement about analyticity of the amplitude under discussion and then extrapolates at *low* energy the asymptotic form of the amplitude. This last step demands some model, usually a Regge parametrization, and a saturation hypothesis, allowing to cut the integrals at low values.

In photon initiated processes one does not need to do that. No model is needed to determine the asymptotic form of the amplitude and to extrapolate it. At fixed ω a point can be reached either by high ν *and* q^2 or both being proportionally smaller. Hence *independent measurements* test the equations directly.

So, no hypothesis about the nature of the J plane singularities controlling the Bjorken limit is really needed. A fixed q^2 sum rule can immediately be written using the hadron techniques to read

$$\frac{2m}{q^2} \int_0^{\nu_m} \nu W_2(\nu, q^2)\, d\nu = \int_1^{1 + W^2/q^2} d\omega'\, F_2(\omega') \tag{19}$$

where we have introduced ω' as defined in the coming section. Here we do face an ambiguity. Though the asymptotic form ω of the variable is fixed, the *variable* that averages the low v, q^2 data is not prescribed a priori. The possibility of using expressions other than ω as a scaling variable was pointed out at SLAC [8] and since then and even before other improvements have been suggested.

If we look at Figs. 3–7 we see the typical behaviour of some data for $q^2 = 1.0$ up 2.5 from Gilman's Tel-Aviv talk.

The resonances very much follow the scaling curve and, what it is striking, is that *each* peak is changing and adjusting itself to the new scaling curve. The scaling variable ω' is given by $(2mv + m^2)/q^2$.

These figures could be drawn for W_1 as well. Traditionally $v\,W_2$ has been preferred because at small angles W_1 could not be measured accurately because of the $\sin^2 \theta/2$ depressing factor.

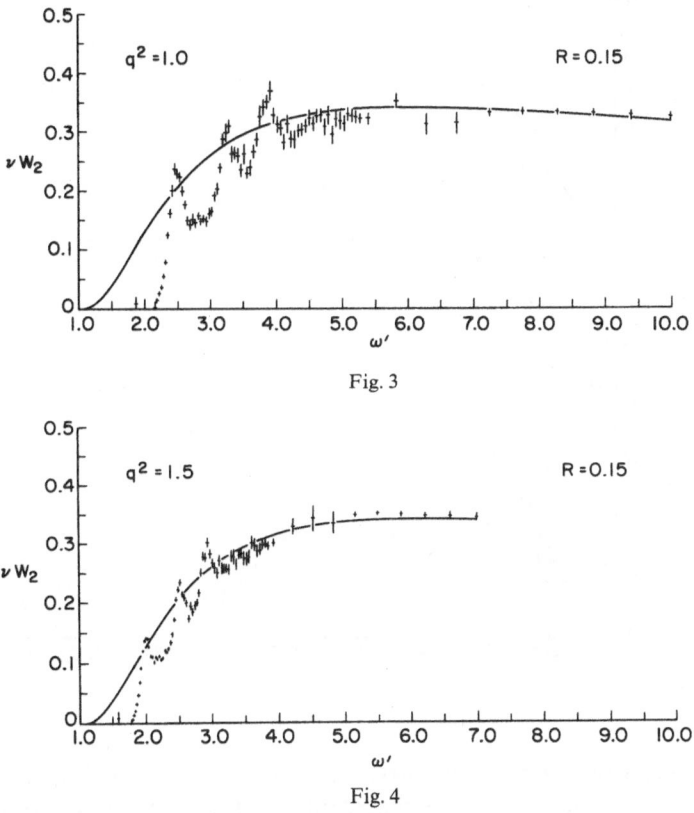

Fig. 3

Fig. 4

Fig. 3–7. $v\,W_2(v, q^2)$ plotted against ω' for fixed q^2 as stated. The solid line is a smooth fit for $q^2 > |W\rangle\, 2\,\mathrm{GeV}$. The arrow indicated the position of the elastic peak (see Ref. [15])

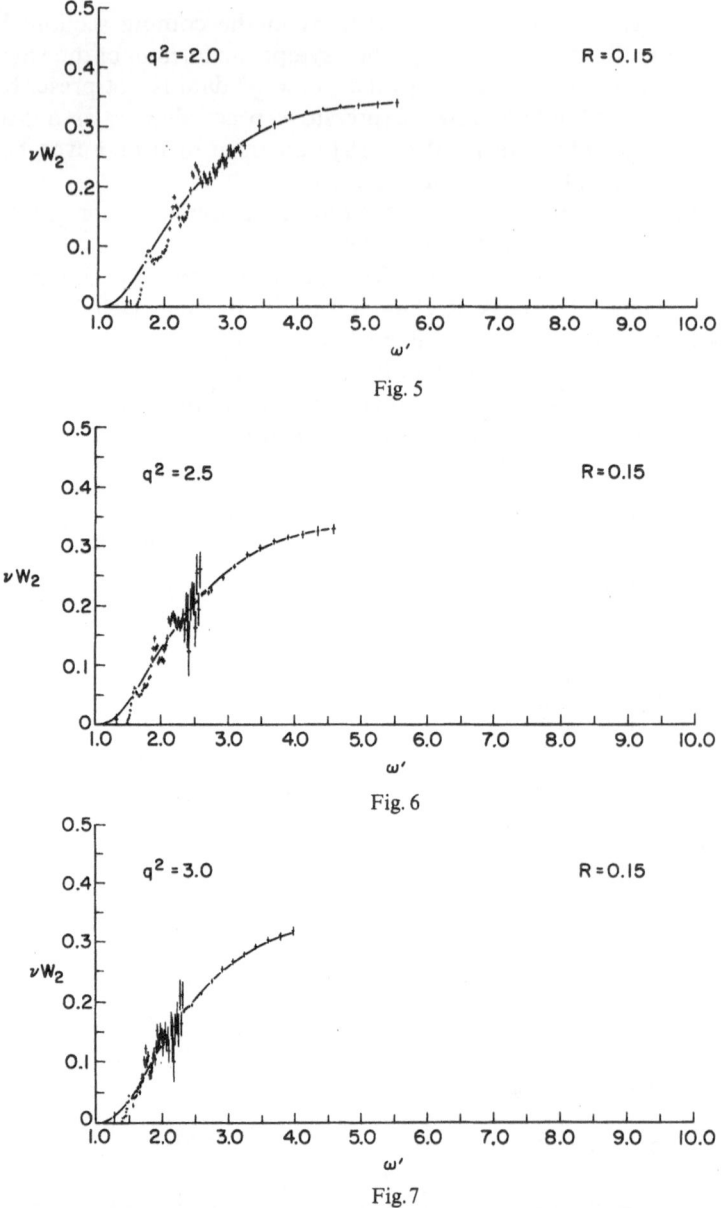

Fig. 5

Fig. 6

Fig. 7

The fact that the scaling curve averages the resonances to 10% from the very beginning is quite remarkable. This is precisely what local duality means: locally the asymptotic form is equivalent to the low energy amplitude.

Historically, early theoretical models [2] conjectured that the resonances could disappear fast when q^2 grows and that the scaling properties witnessed at low q^2 reflected properties of the background or, if we use Regge language, the diffractive Pomeron. We have just seen that this model is ruled out by the data. The agreement between both sides breaks down in the absence of resonances. We will come to that later on again. So the next theoretical question is: can one produce models in which direct channel resonances build a substantial part of the νW_2 function. This is indeed the case and some of these dynamical models [9] are discussed by Landshoff in his lectures.

We want now to discuss the predictions of the local duality model. We warn the reader that strict local duality is of course untenable as resonances peaks do indeed exist. Nevertheless the narrow resonance model is used to show that its predictions are semi-quantitatively correct. The sum rules can be written down to read

$$\frac{2m}{q^2} \int_0^{\nu_m} d\nu\, \nu\, W_2(\nu, q^2) = \int_1^{1 + W^2/q^2} d\omega'\, \nu\, W_2(\omega') . \tag{19}$$

Strict local duality demands that the integrands are roughly equal point by point, or say resonance by resonance. We parametrize the excitation form factors of the system for large values of q^2 as a power law that we assume is universal, then

$$G(q^2) \xrightarrow[q^2 \to \infty]{} (1/q^2)^{n/2} . \tag{20}$$

Near threshold we have that

$$\nu W_2 \to (\omega' - 1)^p \tag{21}$$

since the structure function vanishes when $\omega' \to 1$, dynamically. By evaluating the integral at a resonance locally, and remembering that the contribution to $W_2 \sim \delta(W^2 - m_p^2)\, G^2(q^2)$ we obtain the relation, keeping q^2 large:

$$n = p + 1 . \tag{22}$$

This relation holds quite well experimentally [10]. For the nucleon form factor $n = 4$ and the fall off to threshold of νW_2 seems to be consistent with $p = 3$. Data however does not determine the slope very near threshold, only up to $\omega' \cong 1.5$. The argument implies that the relation holds universally for all resonances. This is also compatible with the data [8]. The corrections one would expect due to finite widths are not easy to estimate but the results seem quite encouraging.

Similar arguments can now be applied to obtain an estimate of the ratio of the neutron and proton structure functions. Since the normalization of these form factors is known one integrates up to about the nucleon pion threshold (W_{thresh}). Taking the derivative of both sides of (19) and keeping always q^2 large one obtains

$$\nu W_2(\omega' = 1 + W_{thresh}^2/q^2) = (\omega' - 1)^{-1}(-q^2 \, d \, G^2(q^2)/d \, q^2) \qquad (23)$$

where $G(q^2)$ is essentially at large q^2 the magnetic form factor.

$$G^2(q^2) \xrightarrow[q^2 \to \infty]{} G_E^2(q^2)/q^2 + G_M^2(q^2) \qquad (24)$$

It immediately follows that [8]

$$\frac{(\nu W_2)_p}{(\nu W_2)_n} \xrightarrow[\omega' \to 1]{} \left(\frac{\mu_n}{\mu_p}\right)^2 = 0.47 \qquad (25)$$

where *one has assumed the same functional form* of the neutron and proton form factors for large q^2. This result is compatible with the present date but one only has values that are not that near threshold. It is interesting that recent parton models [11] have predicted the ratio to be

$$\left(\frac{\mu_n}{\mu_p}\right) = 0.66 \, . \qquad (26)$$

This value is not ruled out but seems not be preferred by experiment.

To conclude this section we should like to emphasize some crucial points that seen by now definitely established.

a) Resonances survive in the large q^2 regime.

b) Though they may constitute the whole νW_2 at small ω', background cannot be ruled out, specially for high ν.

c) At least for fixed, relative large q^2 duality holds locally in the sense that after every resonance the sum rules holds to better than 10% in accuracy and that their presence is essential to get agreement in Eq. (19).

d) If one assumes a Regge picture and this is something unrelated to the previous points, the presence of resonances demands the existence of "normal trajectories" and at most some diffraction. The study of Regge-like models and how they fit is beyond the scope of my lectures. However, I feel that there is much to be desired about their ability to explain the data. We will come back to this point in our last section.

Llewellyn Smith has also pointed out that if local duality holds for pion targets then the Born term predicts a very large (really infinite σ_S/σ_T ratio). This of course contradicts the quark model algebra. I think that if local duality holds there too, the conclusion is inescapable and would be a point where the marriage of partons and duality may irrevocably dissolve.

Duality in the Whole $v—q^2$ Plane

As stated in the introduction varying q^2 one could learn much more than in the purely hadronic case where the external masses are fixed. Since duality is not rigorously local and form factors are rapidly varying, new information can be gained by writing sum rules along lines of the form $av + q^2 = b$ where a and b are constants. Two problems must then be solved. One, is to establish enough analyticity so as to write the relevant sum rules. Second to make some dynamical assumption so that the sum rules gain content. We first discuss the problem of existence.

The forward scattering amplitude for massive Compton scattering (neglecting spin complications) can be written in three different ways

$$T_i = \int e^{iq \cdot x} dx \, \xi_i \langle p| \, [J_a(x), J_b(0)] \, |p\rangle \tag{27}$$

where $\xi_i = \theta(x_0)$ for the retarded function,

$\xi_i = \theta(-x_0)$ for the advanced function,

$\xi_i = T_+$ time ordered product for the scattering function.

All of these agree in the physical region since they differ by terms that vanish when the four momenta are physical. This is so due to the stability of the one nucleon state. However, whenever an analytic continuation for complex values of v and eventually q^2 is performed the results will be, of course, different.

To study analyticity one should use some representation that explicitates singularities. We use for this purpose the Deser-Gilbert-Sudarshan representation [12] (DGS). This representation is based on microcausality and some technical assumptions about the particular model under discussion. Nakanishi has proven that every connected Feynman diagram satisfies a DGS representation, and these Feynman diagrams are in turn connected to the expression (27). Hence we are able to establish results to all orders in perturbation theory. This is perhaps not enough because of what we said before about the Bjorken limit [4] but it is the best we can do. A derivation of the equations can be found in the Boulder Lectures by Brown. We simply quote the results and outline the procedure to study the analytic properties. The retarded or advanced functions read:

$$T_{\substack{R\\A}} = \int\limits_{-1}^{1} d\beta \int\limits_{0}^{\infty} d\sigma \, \frac{h(\sigma, \beta)}{(-q^2 + \beta - \sigma \pm i\varepsilon(v/2 + \beta)} \tag{28}$$

where we have set $M = 1$ and $h(\sigma, \beta)$ is a weight function whose support is established from the spectral conditions from the commutator.

A slightly different representation holds for T,

$$T = \int\limits_{-1}^{1} d\beta \int\limits_{0}^{\infty} \frac{d\sigma \, h(\sigma, \beta)}{(-q^2 + \beta - \sigma + i\varepsilon)} \tag{29}$$

so that

$$-\pi I_m T = \int\limits_{-1}^{1} d\beta \int\limits_{0}^{\infty} d\sigma \, \delta(\beta - \sigma - q^2) \, h(\beta, \sigma). \tag{30}$$

Hence the problem is to see when the line

$$-q^2 + \beta - \sigma = 0 \tag{31}$$

intersects the support of $h(\sigma, \beta)$. For v and or q^2 fixed the analyticity is easily established. If the mass q^2 is timelike then after the first threshold the v axis becomes fully cut. Real analyticity is lost and some procedure of analytical continuation must be proposed. The new function will not agree either with T_A or T_R. If the vectors q and p are real the support in the physical region of $h(\sigma, \beta)$ is related to a parabola $v^2 = 4q^2$. Any line $av + b = q^2$ *not* entering the region bounded by this parabola has a gap in the v axis and hence a standard finite energy sum rule can be written. Otherwise the problem of continuation must be faced. Of course these are the most pessimistic possibilities. Models need not have these singularities. The generalization to lines that avoid the parabola for lines of the form $av + b = q^2$ was first proposed by Leutwyler and Stern [13].

However, in general the sum rules can be written and they will read

$$\int\limits_{v_a}^{\infty} [v \, W_2(av + b = q^2) - f(\omega)] \, dv + \int\limits_{-\infty}^{v_b} (v \, W_2 - f(\omega)) \, dv$$

$$+ \int\limits_{\text{complex cuts}} [v \, W_2 - f(\omega)] \, dv = 0 \tag{32}$$

where the last term reflects the possibility of having to integrate over complex cuts.

Fortunately we need not worry because we add now the hypothesis of semi-local duality [14]. This dynamical assumption states that these integrals must vanish by themselves since the amplitude equals *locally* its asymptotic part. Hence, we have established that, if semilocal duality holds, finite energy sum rules are expected to exist in any direction in the $v - q^2$ plane.

Now we want to check whether these new equations check experimentally. The sum rules originally tested by Bloom and Gilman were the fixed angle sum rules. These read

$$\int\limits_{v_0}^{v_m} \frac{(v \, W_2(\omega_W) - f(\omega_W))}{(2m + 4E\sin^2\theta/2\omega_W)} (q^2 + a^2) \, dv. \tag{32a}$$

Hence, a logical starting point, partly determined by the available data is to choose a direction perpendicular to a small angle. This way a fixed ω line of about 2–3 is the best choice.

We combined the data of a large number of experiments since these are performed at either fixed missing mass or fixed angle. After processing the data to reconvert it for our purposes we plotted $\widetilde{\nu W_2} = \nu W_2/f(\omega')$ as the ordinate against a variable roughly proportional to ν. It reads

$$\tilde{\nu} = (\nu + m/2)\,(\omega'-1)/\omega' = W^2/2m. \tag{33}$$

This variable has the advantage that whatever ω' the resonances appear at the same value of $\tilde{\nu}$. This is needed to test properly local duality.

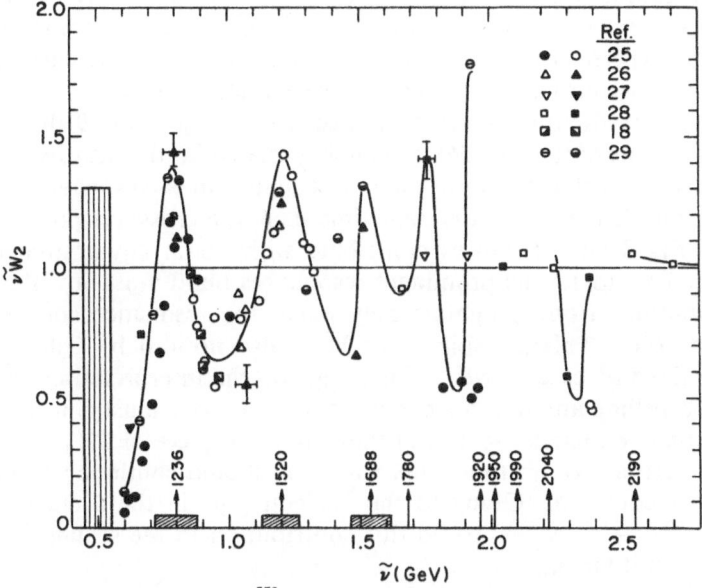

Fig. 8. Fixed ω' sum rule. $\widetilde{\nu W_2} = \dfrac{\nu W_2}{f(\omega')}$, is defined by Eq. (3). The origin of the experimental points is given in the upper corner of the figure. The first column corresponds to $2 < \omega' < 2.5$ the second one $2.5 < \omega' < 3$. The curve is traced by eye. Representative errors are shown for two points. The position of the resonances is shown; for the first three of them their width is given

Again to compare points with different ω' we divided νW_2 by the value of its asymptotic form. For this region of ω' the variations of the asymptotic form are still considerable. This way all points are plotted against a universal function. Notice that we choose ω' the same variable as in the previous section. We will show later on that there is a better one ω_W but they only differ for small q^2. Hence, to avoid ambiguities we restricted ourselves here for large values of q^2. The acceptance $\varDelta\omega'$ was rather large to be able to gather enough data. We did separate the points between 2 and 2.5 and 2.5 and 3 but no systematic variation was observed. We finally plotted all points together. By looking at Fig. 8

we first observe that the local average obtains to a very good approximation when we plot the resonances against ω'. ω *does not* average the resonances in this direction either.

If we cut the sum rule at the bottom of the valley after every resonance we find that both sides are equal to about 10%.

The contribution of the nucleon pole was computed explicitly. However, its local contribution is undefined and we arbitrarily spread it over a typical hadron width. The curve was traced by eye aided by the known resonances widths and positions. We have no way to estimate errors in a reliable fashion and hence we took seriously the ones quoted in each experiment. Under these circumstances the agreement seems beyond expectations. There are several qualitative aspects of the picture that deserve attention. First we notice that in Fig. 8 and 9 the relative contributions of the $\Delta(1236)$ is changing relative to the next resonances. This indicates that the form factors of higher mass resonances have a higher tail than Δ but at the same time there is a crossover point.

In Fig. 8 all resonances contribute with about equal strength. In fact for $\tilde{v}$ up to 1.5 the prominent resonances build most of $v W_2$. Small contributions from daughters may make the resonance contribution almost 100%. This possibility has been also noticed by Gilman [15] in the fixed q^2 case as well. When $\tilde{v}$ grows the interpretation becomes less compelling and nothing can be said. Hence one must conclude here again that resonances scale and they are a big piece of $v W_2$.

To further exploit those sum rules one should continue and use the left over equations. Going to the nucleon pole in the u channel and fixing $m_\varrho^2 = -q^2$ we compared this contribution to the scaling function and found it far off.

This result is perhaps not surprising since there is no reason to expect the mass singularities to be dual to the intermediate hadronic states. This may be one reason why models for currents including duality are still facing difficulties. One way to see how individual resonances may scale in theories (dual models) where there is a growing caracteristic mass is the following. One worries about having large masses in the theory since the trajectories raise continuously. However, if the current couples through vector mesons *also* through members of these trajectories the universal slope is the same in the current *and* t channels and the scaling obtains as the slope cancels [16].

Extension of Local Duality as a Function of q^2

We want now to extend duality and scaling up to $q^2 = 0$. If we look at Fig. 9 we see that the averaging performed by ω' is not very good for low values of q^2 for $v W_2$ *or* the experimentally measured cross-section.

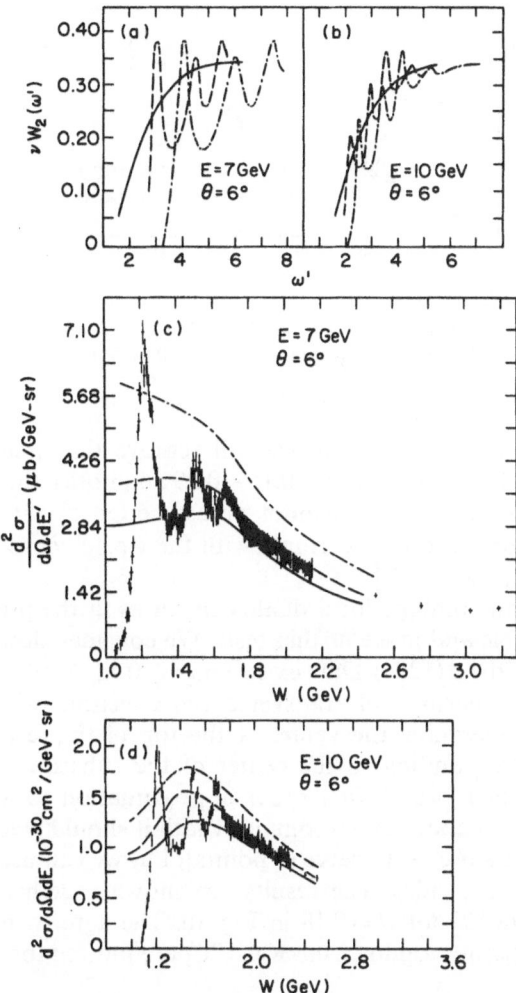

Fig. 9. a, b plot of νW_2 versus both against ω' (dashed dotted line) and ω_w for $a^2 = 0.15$ (dashed line). The smooth averaging curve corresponds to high q^2 and ν data assuming $R = 0$.

c, d plot of $\dfrac{d^2\sigma}{d\Omega' dE'}$ against W. Data from Ref. [13]. The averaging corresponds respectively, going upwards, to $a^2 = 0.25$; 0.15 and zero

Because of kinematical factors and the existence of both longitudinal and transverse contributions the two tests are not equivalent. So any averaging variable should affect only low q^2 values. We propose

$$\omega_w = \frac{2m\nu + m^2}{q^2 + a^2} = \frac{s + q^2}{q^2 + a^2}. \tag{34}$$

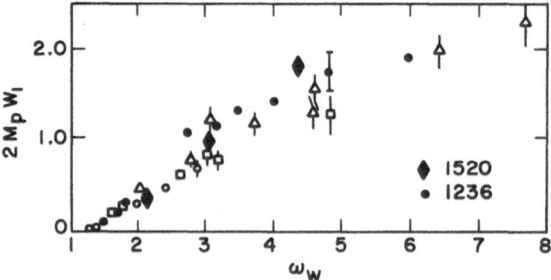

Fig. 10. $2mW_1$ plotted against ω_W. Data from Ref. [13]. Circles are points calculated from data from Ref. [5]. Rhomboids correspond to the 1520 MeV resonance, Ref. [7]

We first assume $R = \sigma_S/\sigma_T = 0$ (we will remove this assumption later) and check local duality with $a^2 = 0.15 - 0.25$ and notice that local duality is much better verified. We cannot adjust better a^2 with the data we had but an effort to test the formula with the world compilation of data is already under way [17].

Now we can combine local duality much as in the previous section with our variable and make further tests. We consider electroproduction in the region of the Δ (1236). Data exists for $q^2 \geq 0.1$ [18, 19]. We computed W_1 from the experimental transverse cross-section, which is known separately by averaging the values at the top of the resonance and at $W = 1390$, corresponding to the center of the subsequent dip. This is of course a rather brutal, first order approximation to averaging, but if duality holds to about one resonance width it should give a reasonable answer. The data are very scarce (6 points!) but we can use *both* fixed q^2 and fixed ω_W sum rules. The results are shown together with the experimental data [7] for $R = 0.18$ in Fig. 10. The agreement is really excellent. A few points coming from Ref. [20] are plotted too and they also fit well.

We next compute $R = \langle \sigma_S \rangle / \langle \sigma_T \rangle$. We assume that the ratio itself is smooth. Values of the order of 0.2 are obtained from the Δ(1236) including points with $q^2 = 0.1$. Unfortunately errors are too big to allow for a meaningful determination of R as a function of ω_W. It is again no small triumph of duality that the resonance value of R is the same as obtained at high energies and high q^2.

The next excercise is to discuss νW_2. Here we have as shown in Eq. (10) a kinematical zero. Hence we must modify the structure function to eliminate this effect. The simplest way to eliminate this "threshold effect" is to consider [14]

$$(\nu W_2)_{\text{th}} = \omega/\omega_W (\nu W_2). \qquad (35)$$

We can test our variable against the SLAC data and see whether the threshold change of νW_2 as a function of ω as explained. The agreement is excellent. In fact we predict that for

$$\omega = m^2/a^2 \qquad (36)$$

the slope will change sign. This seems to be indeed true. The results are shown in Fig. 11. Unfortunately, there is some averaging over ω that precludes an even stronger test, but the ω dependence of the slope is beautifully explained. We have found that properly interpreted the

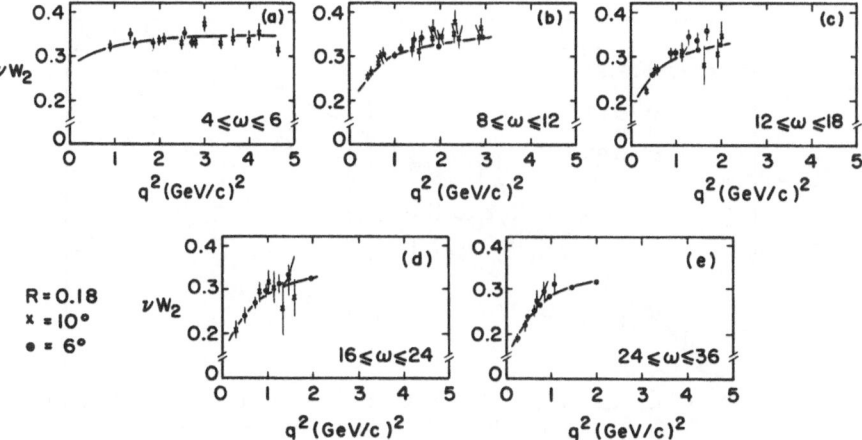

Fig. 11. νW_2 behaviour as a function of q^2 for fixed averaged values of ω. From Ref. [7]

whole electroproduction amplitude is a function of one variable throughout the whole ν and q^2 range. The Δ resonance contains all the information about electroproduction. This is not completely true since there is a bound on ω_W of about 6. This makes the connection with diffraction satisfactory since at the Δ the background is small and the high ν limit seems diffractive in nature.

Our connection between duality and scaling seems to hold. It is a very strong restriction on theories. Available dual resonance models do not seem to have the local averaging variable as demanded by the data. By far the most intriguing aspect of the problem is the ability of the Δ resonance alone to reproduce the asymptotic scaling function. Narrow resonance models in which other trajectories are to be included as additive terms have no contribution to the imaginary part at the Δ position. Hence, it might indicate that one needs more complicated than additive models to explain the data, since the Δ by itself by varying

q^2 reproduces the *whole* structure function to which these other trajectories contribute as well. Hence duality seems here even stronger than in the strongly interacting case.

Connection with Photoproduction

In the previous section we have extended duality to $q^2 = 0.1$. One can immediately ask if one can take the limit $q^2 = 0$ and make a fruitful connection with photoproduction.

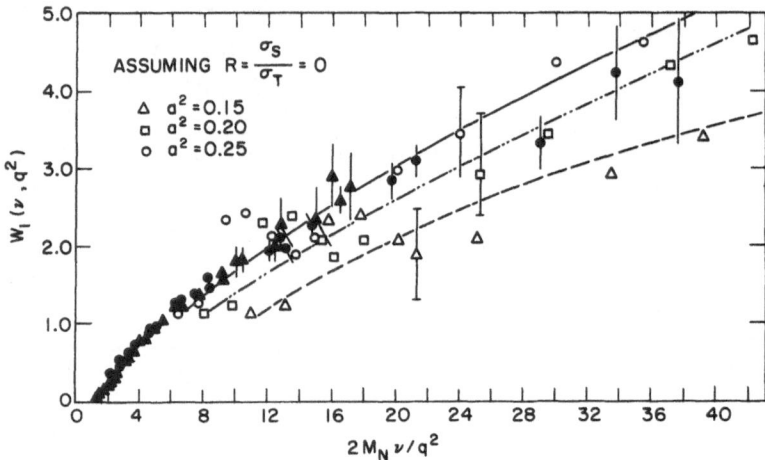

Fig. 12. $2m W_1$ plotted against ω_W. Data from Ref. [13]. The curves are calculated from photoproduction data, Ref. [14], for $a^2 = 0.15$, 0.20 and 0.25 respectively starting below

To that effect we can simply compute the structure function W_1 by taking the total cross section data [21] and going by the same technique of averaging from top to bottom of each resonance. The results are satisfactory but very sensitive to the value chosen for a^2 (see Fig. 12).

The possibility of using real photon data allows us to make some striking predictions concerning the neutron proton structure function difference.

The argument is very simple. The total inelastic cross-section for γ rays on protons or neutrons near threshold is the 1 pion production one. At threshold one has a low energy theorem, the Kroll-Ruderman theorem, that states that

$$\sigma_{\gamma p \to p \pi^0} = \sigma_{\gamma n \to n \pi^0 = 0} \, , \tag{37}$$

$$\sigma_{\gamma p \to n \pi^+} = \sigma_{\gamma n \to p \pi^-} \, . \tag{38}$$

Hence the total cross-sections are equal. Using formula (8) or (35) if one wants νW_2 we predict that around that value the difference must vanish. Of course, duality is not strictly local but a detailed study of the multipole models that fit the data very well [22] demand in fact that throughout the first few hundred MeV $\sigma_{\gamma n} \gtrapprox \sigma_{\gamma p}$. Hence we predict that the difference between neutron and proton must vanish or become slightly negative at the point $\omega_W \sim 6$–7, where $\bar{\nu}$ is some value around the first resonance.

Though the analysis is slightly model dependent if one accepts the predictions, at the second resonance the proton has taken over again. See the accompanying Fig. 13 [23].

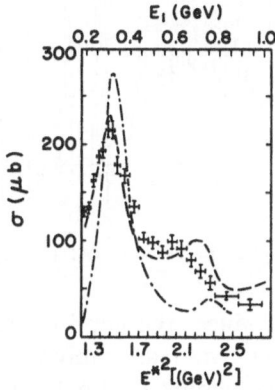

Fig. 13. Total cross section for the reaction $\gamma + n \rightarrow p + \pi^0$ as a function of incoming energy given by $++++$, $\gamma + p \rightarrow n + \pi^+$ given by $--$, $\gamma + p \rightarrow p + \pi^0$ given by $-\cdot-\cdot-$ (see Ref. [23])

The value here for ω_W is about 8. For large values of ω there are data [21] and the difference between the proton and neutron does tend to zero.

If these oscillations really take place, then the usefulness of the Regge parametrization at moderate values of ω can be put in doubt.

Though it is too early to decide because of problems with the data, more tolerable errors in this difference will be of importance.

If one goes to electroproduction with small q^2 at the $\Delta(1236)$ resonance, one moves to smaller ω_W and hence one expects, if the theory works, that the proton cross sections will grow, relative to the neutron one, with q^2. There may be some evidence already in favour of this prediction.

The most puzzling aspect of this work is the origin of a^2. A detailed study of the data may allow for a^2 slightly bigger, if one allows the m^2 parameter in the numerator to change as well. It is perhaps no accident that both constants are roughly the two numbers that seem to characterize strong interactions: the transverse momentum length and the universal slope of Regge trajectories.

From the theoretical point of view the breaking of scale invariance seems to be characterized by a small number a^2. There have been some attempts [24] to relate by use of Regge theory photoproduction and inelastic electron scattering. This model does not predict the proper

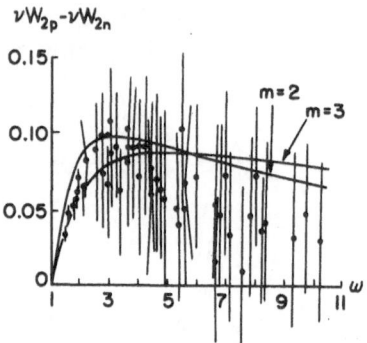

Fig. 14. $\nu W_{2p} - \nu W_{2n}$ as a function of ω. From Ref. [7]. Typical Regge fits from Ref. [9]

scaling variable but introduces the good a^2 dependence of the residue function $\beta \sim (q^2 + a^2)^{-1}$ that gives reasonably well the threshold behaviour. However local duality is missing in this model.

It will be fascinating to see, if the marriage of duality and scaling will produce a theory with some success.

Both concepts came from high ν and q^2 speculations, but seem to apply to embarrassingly low values of these variables. We hope that this uncertain situation will stimulate further theoretical work.

Postscriptum, November 1971

Since the lectures were delivered, the work reported in Ref. [17] has been finished. It lends strong support to ω_W as a scaling variable. The parameters are slightly changed and the solutions though still not conclusive seem to favor slightly larger a and m.

Also, at Cornell, the ratio of structure functions for protons and neutrons is reported to be even smaller than the one predicted by (25). This is no blow to duality since the additional assumption, as explained in the text, of form factor scaling was used. In fact this assumption is also known to fail. The situation with the difference in structure functions is unfortunately still confused after the Cornell Conference and awaits further experiments.

References

1. See for example the accompanying lectures of Llewellyn Smith and references therein.
2. Harari, H.: Phys. Rev. Letters **22**, 326 (1969). – Abarbanel, H., Goldberger, M., Treiman, S.: Phys. Rev. Letters **22**, 500 (1969).

3. For a discussion of these points see any recent papers on Compton Scattering, e.g. Kugler, M., Milgrom, M.: Nucl. Phys. B, to appear.
4. See, for example, the work of Adler, S., and others.
5. Altarelli, G., Rubinstein, H. R.: Phys. Rev. **187**, 2111 (1969).
6. Frishman, Y.: Amos de Shalit Memorial Volume Contribution, Ann. Phys. 1971.
7. See for example the contributed paper by Bloom, E., et al. at the 1970 Kiew Conference.
8. Bloom, E., Gilman, F.: Phys. Rev. Letters **25**, 1140 (1970).
9. Domokos, G., Koveri-Domokos, S.: John Hopkins University Preprints (1970). – Landshoff, P. V., Polkinghorne, J.: Nucl. Phys. B **19**, 432 (1970) and lectures of one of the authors in this volume.
10. This relation was first proposed by Drell, S., Yan, T. M.: Phys. Rev. Letters **24**, 181 (1970). – West, G.: Phys. Rev. Letters **24**, 1206 (1970).
11. Weisskopf, V.: Invited talk at the 1971 Trieste Symposium and Landshoff, P., Polkinghorne, J.
12. Most of the results of this section were established in collaboration with Frishman, Y., Rittenberg, V., to be published. In the meantime an excellent paper on the subject by Ashok Suri SLAC preprint 738 appeared. It contains a detailed discussion of methods needed to establish analyticity in the two complex variable problem.
13. Leutwyler, H., Stern, J.: Phys. Letters **31** B, 458 (1970).
14. Rittenberg, V., Rubinstein, H. R.: Phys. Letters **32** B, 50 (1971).
15. Gilman, F.: Invited talk at the Tel Aviv Conference 1971 SLAC preprint 896.
16. Rittenberg, V., Rubinstein, H. R.: Nucl. Phys. B **28**, 184 (1971).
17. This work is being conducted at DESY. Nuclear Phys. to appear.
18. Bartel, W., et al.: Phys. Letters **27** B, 225 (1968).
19. Lynch, H. L., et al.: Phys. Rev. **164**, 1635 (1967).
20. Albrecht, W., et al.: Nucl. Phys. B **13**, 1 (1969).
21. Joos, P.: Compilation of photoproduction data above 1.2 DESY-Hera 70–1.
22. See for example the recent Technical Report by Schwela, R.: Bonn preprint 1971. Of course, changes due to the possible existence of an isotensor term may change the effect slightly, but the abrupt decrease seems inevitable.
23. Rizzini, E. Lodi, et al.: Nuovo Cimento Lettere **3**, 697 (1970). The arguments depends on the neutron cross-section predictions from Schwela.
24. Moffat, J., Snell, V.: Phys. Rev. D **3**, 2848 (1971).
25. Core, A. A., et al.: Phys. Rev. **164**, 1490 (1967).
26. Brasse, F. W., et al.: Nuovo Cimento **55** A, 679 (1968).
27. Albrecht, W., et al.: Phys. Letters B, 225 (1968).
28. Bloom, E., et al.: Phys. Rev. Letters **23**, 930 (1969). – Breitenbach, M., et al.: Phys. Rev. Letters **23**, 935 (1969).
29. Albrecht, W., et al.: DESY preprint 69/46.

Dr. Hector R. Rubinstein
Department of Nuclear Physics
Weizmann Institute of Science
Rehovot/Israel

Scaling in Deep Inelastic Scattering with Fixed Final States* **

V. Rittenberg

Contents

1. Introduction

Operator product expansions in the vicinity of the light cone [1] have been proven to be a useful tool to study processes involving high virtual mass photons. However, with the exception of the well known case of deep inelastic electron photon scattering, supplementary assumptions are necessary.

In the present lecture we will show that further tests of these ideas can be made for the electroproduction of a final state as well. Here too some supplementary assumptions will be used. We consider that the general structure obtained in the Björken limit for these processes will provide a stringent test of the validity of the operator expansions.

Our result is that for the one particle electroproduction

$$\text{"}\gamma\text{"} + A \rightarrow B + C \tag{1}$$

the invariant amplitudes A_i can be expressed in the Björken limit $(v \rightarrow \infty, q^2 \rightarrow -\infty; \omega$ and t fixed) as:

$$A_i(v, q^2, t) \sim v^{d_i} f_i(\omega, t) \tag{2}$$

where $t = (q - p^b)^2$, $\omega = 2M_A v/-q^2$, $v = p^A q/M_A$, q is the photon four momentum. The fact that the d_i are independent of ω and t is a test of the conjecture that states that the singularities of the operator expansions near the light cone are c-numbers. As will be discussed below the power of our method may be increased if we conjecture that the numbers d_i

* Lecture delivered at the International Summer Institute in Theoretical Physics, DESY, July 12–24, 1971.

** Work supported in part by Deutscher Akademischer Austausch-Dienst grant.

may be obtained assuming naive dimensionality for the operator products. The same type of analysis can be applied for many particles electroproduction:

$$\text{``}\gamma\text{''} + A \to B + C_1 + C_2 + - + C_n \tag{3}$$

if the particles C_i are fragments of A and B is a fragment of "γ" and henceforth to the inclusive process

$$\text{``}\gamma\text{''} + A \to B + \text{anything}$$

if B is a fragment of "γ" and we are in the fixed missing mass Björken limit.

2. Application of Light Cone Expansion for one Particle Electroproduction

We illustrate the method by studying the scattering of scalar photons on scalar equal mass isospinless particles

$$\text{``}\gamma\text{''}(q) + A(p^A) \to B(p^B) + C(p^C). \tag{4}$$

Straightford application of the reduction formalism allows us to write the amplitude as

$$T \sim \int dx\, e^{iqx} \theta(-x_0) \langle A | [j(x), j_B(0)] | C \rangle \tag{5}$$

where $j(x)$ is the "electromagnetic current" and j_B is the source of the scalar field. We have omitted possible terms containing equal time commutators, a procedure that we justify later. The approximation scheme of the expansion demand that one finds a frame in which light cone dominance can be proven. We choose the rest frame of particle A. In this frame:

$$q = (q_0, 0, 0, q_3); \qquad p^B = (p_0^B, 0, p_2^B, p_3^B)$$
$$p^A = (M, 0, 0, 0); \qquad p^C = (p_0^C, 0, p_2^C, p_3^C). \tag{6}$$

We consider the limit $\nu \to \infty$ and ω, p_2^C and p_3^C fixed (particle C is a fragment of A). This corresponds to fixed t limit since

$$|p^C| = (1/2M)\, [t(t - 4M^2)]^{\frac{1}{2}},$$
$$\cos\theta_C = (2M^2 - \omega t)/2M\omega |p^C| \quad (p_3^C = |p^C| \cos\theta_C). \tag{7}$$

On invariance grounds, the matrix element $\langle A | [j(x), j_B(0)] | C \rangle$ can depend on $x^2, x p^A, x p^C$ and t which are all independent of ν. Hence all the ν dependence is isolated is the exponential e^{iqx} since $q_0 = \nu$, $q_3 \simeq \nu + M/\omega$.

Now

$$\exp iqx = \exp i\nu(x_0 - x_3) + iM x_3/\omega.$$

The integral picks up contributions from $|x_0 - x_3| \sim 1/v$ and $|x_0 + x_3| \lesssim \omega/M$. Hence $|x_0^2 - x_3^2| \lesssim \omega/Mv$. From causality we know that the matrix element is non zero only for $(x_0^2 - x_3^2) - (x_1^2 + x_2^2) \geq 0$ so that we have finally $x^2 \lesssim \omega/Mv \to 0$ and hence light cone dominance. We now apply the light cone expansion of operator products:

$$\theta(-x_0) [j(x), j_B(0)] = C(x) F(x) + \cdots \tag{8}$$

where

$$C(x) = \theta(-x_0)|(-x^2 + i\varepsilon x_0)^d - (-x^2 - i\varepsilon x_0)^d| \tag{9}$$

and $F(x)$ is an operator which is nonsingular for $x^2 = 0$. Inserting (8) in (5) we get:

$$T \simeq \int dx\, e^{iqx} C(x) F(xp^A, xp^C, t) \tag{10}$$

where $F(xp^A, xp^C, t)$ is the matrix element of the nonsingular operator F. In order to do the integrations in x we consider the double Fourier transform of the matrix element:

$$F(xp^A, xp^C, t) = \int_{-\infty}^{1} \int_{-1}^{\infty} d\alpha\, d\beta\, g(\alpha, \beta, t) \exp i(\alpha p^A + \beta p^C)\, x. \tag{11}$$

The limits of integration in (11) are obtained from the spectral conditions. (In writing (11) we have assumed that there are no end point singularities.)

We use now the identity

$$\int dx\, e^{ikx} C(x) = C(d)\, h(k^2, k_0) \tag{12}$$

where

$$C(d) = \pi 2^{2d+3} \Gamma(d+1)\, \Gamma(d+2)$$

$$h(a, b) = e^{-i\pi\varepsilon(b)}[(-a + i\varepsilon)^{-d-2} - (-a - i\varepsilon)^{-d-2}] \tag{13}$$

$$+ (a + i\varepsilon)^{-d-2} - (a - i\varepsilon)^{-d-2}$$

and get

$$T \simeq C(d) \int_{-\infty}^{1} \int_{-1}^{\infty} d\alpha\, d\beta\, g(\alpha, \beta, t) h(\xi^2, \xi_0) \tag{14}$$

where

$$\xi_0 = (q + \alpha p^A + \beta p^C)_0$$

$$\xi^2 = (q + \alpha p^A + \beta p^C)^2.$$

This is all what light cone expansions give us. In order to prove scaling we shall now assume that only finite α and β contribute to the integrals in (14). In this case

$$\xi_0 \simeq v$$
$$\xi^2 \simeq 2Mv[\alpha + \beta(1 - 1/\omega) - 1/\omega]. \tag{15}$$

A similar assumption was not necessary in proving scaling from light cone dominance for the deep inelastic total electroproduction cross section. However, it can be shown that at least in some models our assumptions are valid [4]. In fact the same type of assumption is needed in order to prove light cone dominance in mass dispersion relations for form factors [5].

From (14) and (15) we get finally

$$T \sim v^{-d-2} f(\omega, t) \tag{16}$$

where d is independent of ω and t and depends only on the asymptotic dimension of the commutator of the two considered currents (one electromagnetic and the other one hadronic). The expression (16) gives the scaling behaviour for one particle electroproduction in the fixed t, Björken limit.

The fact that in the Björken limit the real and the imaginary part of the amplitude behave in the same way proves that there is no need for contact terms and this justify the omission of terms containing equal time commutators in the reduction formula (5).

An application of these ideas to pion-electroproduction may be found in Ref. [2].

It is very important to mention that the scaling behaviour (16) was also obtained in the ladder approximation [6] of the ϕ^3 theory using the method of Altarelli and Rubinstein [7]. Since this model admits a light cone expansion [4] this leads support to our form (16).

A different behaviour of the amplitude in the Björken limit is obtained in a dual resonance model for currents [8]:

$$T \sim v^{\xi(t)} f(\omega, t) \tag{17}$$

where $\xi(t)$ (independent of ω!) is related to the Regge trajectories included in the model. May be that there is some incompatibility between light cone expansions and usual dual models but at the present stage it is very difficult to decide since on one side supplementary assumptions were necessary in our derivation and on the other side dual models for currents have still their own problems. It is worthwhile to mention Müller and Rühl's conjecture which also leads to the Regge type behaviour (17).

3. Generalization for many Particles Production

The same procedure may now be applied for the process

$$\text{``}\gamma\text{''} + A \to B + C_1 + \cdots + C_n$$

for fixed $t = (q - p^B)^2$ and the particles $C_1, \ldots, C_n$ being fragments of particle A. Now

$$T_{n+1} \sim \int dx\, e^{iqx} \theta(-x_0) \langle A | [j(x), j_B(0)] | C_1, \ldots, C_n \rangle . \tag{18}$$

Considering again the rest frame of particle A. the particles $C_1, ..., C_n$ having finite momenta, the matrix element will not depend on v and again the whole v dependence is isolated in the exponential. We get

$$T_{n+1} \sim v^{-d-2} f(\omega, t, p^{C_i} p^{C_j} ...) \tag{19}$$

where d is the same like in the one particle electroproduction due to the fact that d is related only to the light cone properties of the commutator of the same currents.

The fact that the exponent d is independent of how many particles are electroproduced, enables us to consider the inclusive process

$$\text{``}\gamma\text{''} + A \rightarrow B + \text{anything}$$

The differential cross section of this process is a function of 4 invariants:

$$p_0^B d\sigma/d p^B = f(v, q^2, \mathcal{M}^2, t) \tag{20}$$

where

$$\mathcal{M}^2 = (q + p^A - p^B)^2; \quad v = q p^A / M.$$

We now consider the Björken limit in the fixed missing mass limit. We shall prove that in the limit $v \rightarrow \infty, q^2 \rightarrow -\infty, \mathcal{M}^2, \omega, t$ fixed we have

$$f(v, q^2, \mathcal{M}^2, t) \simeq v^{d'} f(\omega, \mathcal{M}^2, t). \tag{21}$$

In fact

$$p_0^B d\sigma/d p^B \sim \sum_n \int \frac{d p^{C_1}}{p_0^{C_1}} \cdots \frac{d p^{C_n}}{p_0^{C_n}} \frac{|T_{n+1}|^2}{\lambda^{\frac{1}{2}}} \delta \left(p_A + q - p_B - \sum_{i=1}^n p^{C_i} \right) \tag{22}$$

where

$$\lambda = s^2 + M^4 + q^4 + 2M^2 q^2 + 2s q^2 - 2M^2 s$$

but in the fixed missing mass limit we have (we are again in the rest frame of particle A):

$$\left(\sum_{i=1}^n p^{C_i} \right)^2 = \mathcal{M}^2 \left(p^A - \sum_{i=1}^n p^{C_i} \right)^2 = t \tag{23}$$

and hence

$$\sum_{i=1}^n p_0^{C_i} = \frac{1}{2M} (M^2 + \mathcal{M}^2 - t) \tag{24}$$

which means that all the 3 moments of the particles C_i remain finite in the Björken limit. Thus we can apply our standard technique to get the asymptotic behaviour (19) and from (22) we finally obtain the scaling rule (21) with $d' = -2d - 5$.

All these derivations have included a first step in which the right reference frame and the proper reduction formula has been used in order to get the light cone dominance. However light cone expansions are just in one variable (x^2) and can be used just to do one integration.

This was enough for the forward Compton amplitude in order to prove scaling but this is not the case in general. Thus the next step consists in either making models for the remaining unknown function or, as we did, to make some general assumptions which allowed us to prove scaling. As concerning the first step let us remark that instead of using the reduction formula (5) we could use

$$T \sim \int dx\ e^{-ip^Bx}\theta(x_0)\langle A|[j(0), j_B(x)]|C\rangle. \qquad (25)$$

But in this case $p_B^2 = M_B^2$ is finite so that the proof does not hold. In fact the matrix element is v dependent and we are unable in this way to "see" the light cone.

We have been able to prove scaling like behaviour only for the differential cross section of exclusive processes in certain kinematical regions. This does not mean that the *integrated* cross sections should scale and a priori we have room essentially for any kind of behaviour for the integrated cross sections. Although from some models scaling has been conjectured also for these quantities [11], such models seem to have no obvious connection with light cone dominance.

During the preparation of this talk I have enjoyed discussions with R. A. Brandt, G. Preparata, B. Renner and H. R. Rubinstein.

References and Footnotes

1. Frishman, Y.: Phys. Rev. Letters **25**, 966 (1970); Altarelli, G., Brandt, R. A., Preparata, G.: Phys. Rev. Letters **26**, 42 (1971).
2. — Rittenberg, V., Rubinstein, H. R., Yankielowicz, S.: Phys. Rev. Letters **26**, 768 (1971). Frishman, Y.: Talk at the 1971 Coral Gables Conference.
3. Georgelin, Stern, J., Jersák, J.: Nucl. Phys. B**27**, 493 (1971) have first integrated in $(x_0 + x_3)$ and afterwards looked to the $(x_0 - x_3) \to 0$ limit. In that way they got scaling only for $M_B^2 \leqq 0$.
4. Brandt, R. A., Preparata, G.: (privare communication).
5. Preparata, G.: Lecture delivered at the International Summer Institute in Theoretical Physics, DESY 1971.
6. Matsuda, S., Suzuki, M.: Phys. Rev. D1, 1778 (1970). Dorren, D. J.: Nucl. Phys. B (to be published).
7. Altarelli, G., Rubinstein, H. R.: Phys. Rev. **187**, 2111 (1968).
8. Drummond, I. T.: CERN preprint 1971.
9. Müller, V. F., Rühl, W.: Nucl. Phys. B (to be published).
10. Rubinstein, H. R., Veneziano, G., Virasoro, M.: Phys. Rev. **167** 1441 (1968).
11. Lee, T. D.: Columbia University preprint 1971.

Dr. V. Rittenberg
Department of Physics
Weizmann Institute
Rehovoth/Israel

Duality and the Pion Electromagnetic Form Factor

Kerson Huang

From general principle the pion electromagnetic vertex, as illustrated in Fig. 1, should have the form $e(p_2 - p_1)_\mu F_\pi(t)$, where $t = (p_2 - p_1)^2$. The factor $(p_2 - p_1)_\mu$ insures gauge invariance when the two pions are the mass shell. The invariant function $F_\pi(t)$ is called the pion electromagnetic form factor.

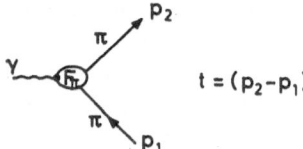

Fig. 1. Pion electromagnetic vertex

The normalization condition $F_\pi(0) = 1$ defines the pion charge to be e.

We have only limited experimental information about the pion form factor. The pion electromagnetic radius has been cited to be [1]

$$r_\pi \equiv \{6[dF_\pi(t)/dt]_{t=0}\}^{\frac{1}{2}} = 0 \cdot 86 \pm 0 \cdot 14 \times 10^{-13} \text{ cm}. \qquad (1)$$

Colliding $e^+ e^-$ beam experiments [2, 3] have measured $F_\pi(t)$ in the neighborhood of the ϱ resonance $t = m_\varrho^2$. When fitted with a Breit-Wigner formular with constant width, these measurements yield a ϱ width of about 100 MeV, which is noticeably different from the width of about 130 MeV that has been obtained from two-pion production in purely hadronic reactions. These measurements, however, are not sufficiently accurate to determine the detailed shape of the ϱ resonance. Indeed they can also be fitted by a squared Breit-Wigner formula [4]. Data on the large $|t|$ behavior, either for $t > 0$ or for $t < 0$, are lacking.

On the theoretical side the situation is equally unsatisfactory. The simplest phenomenological model based on the assumption of vector dominance would give

$$F_\pi(t) = C(t - m_\varrho^2 + i m_\varrho \Gamma_\varrho)^{-1}. \qquad (2)$$

One might add to this contributions from other vector mesons. However, the t dependence of C and Γ_ϱ, which became crucial as one goes away from the resonance position, are unknown. Within the assumption of vector dominance, a theory of $F_\pi(t)$ simply becomes a theory of vector-meson propagators.

If one adopts the view of Lee and Wick [5], namely that propagators are regularized by unstable particles of negative metric, then one could arrive at

$$F_\pi(t) = \int dm^2 \cdot A(m^2)(t - m^2)^{-1}$$
$$+ B[(t - M^2 + iM\gamma)^{-1} + (t - M^2 - iM\gamma)^{-1}], \tag{3}$$

a form explicitly proposed by Schwinger [6] in his theory of sources, in which B is such as to make $F_\pi(t)$ vanish faster than t^{-1} as $t \to \infty$. As noted by Schwinger, the unconventional last term in (3) is permissible because his theory "does not have to contend with the restrictions imposed by an assumed underlying operator field structure, nor with preconceptions about analyticity". The Lee-Wick theory represents a definite way in which the usual "preconceptions" are dispensed with. In the theory of Kroll, Lee and Zumino [7] vector dominance is given a more dynamical role via the concept of "field-current identity". They find

$$F_\pi(t) = F_\varrho(t)/(t - m_\varrho)^2, \tag{4}$$

but the "ϱ form factor" $F_\varrho(t)$ is not calculable within that theory. Although there is no definitive answer about the form of the ϱ pole from these theories, we see that, in the form factor, the ϱ pole may have a more complicated structure than a simple Breit-Wigner formula.

None of the above theories take into account hadron dynamics, or what little we know about it. On the other hand, there have been attempts to construct hadronic currents by emphasizing the duality aspect of hadron dynamics, and then try to satisfy gauge invariance and current algebra requirements [8]. These have not been successful so far, and I shall not go into them.

I would like to explore here a very crude model [9] that combines duality and current algebra requirements in a simple way. Our object lies not so much in producing a formula that can be compared directly with experiment, but rather in exploring the type of qualitative behaviors that could possibly emerge. In this sense it is an experiment in a "theoretical laboratory".

If the pion had no strong interaction, its electromagnetic vertex would be $e(p_2 - p_1)_\mu$, as represented by the first Feynman graph in Fig. 2, and the form factor would be unity. The strong interactions give rise to final state interactions, represented by the second Feynman diagram of Fig. 2, in which A is the pion-pion elastic scattering amplitude. In principle one may have other graphs, in which the two – pion intermediate state is replaced by other hadronic states, namely states containing any number of bosons and baryon-antibaryon pairs of arbitrary spin. However, it does not violate any known principle to make the following model: Only pions pairs are coupled to a photon through the elementary vertex $e(p_2 - p_1)_\mu$. All other charged particles are coupled

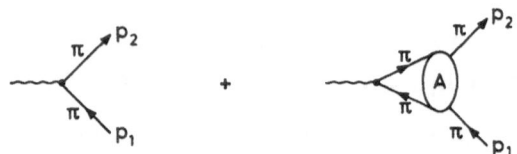

Fig. 2. Our model for the pion form factor

to the photon in a non-elementary way, namely by first becoming a pion pair via the strong interactions. This is our model. If philosophy is important, we might adopt two different interpretations of the model:

(1) The model is a mathematical exercise. We really believe that other hadrons also couple to the photon in an elementary way, but we consider only the pion contribution for the moment.

(2) We believe that the electromagnetic field single out the pion as an "elementary" particle. That is, pions are partons.

I tend to favor the second, if only for the reason that it is less apologetic; but it must be admitted that, if taken literally, this view seems to have been ruled out by experiments. The SLAC-MIT experiments on deep inelastic $e - p$ scattering showed that the virtual photon in that process is predominantly transverse, and this would require the partons to have spin $\frac{1}{2}$ instead of 0. Thus we have to retreat a little and say that we would perhaps like to use spin $\frac{1}{2}$ partons, but for technical reasons (namely, that a dual resonance model for boson-fermion scattering is unknown) we pretend that they have spin 0.

The advantage of introducing electromagnetism in this manner is that gauge invariance and current algebra requirements can be easily satisfied. Gauge invariance is automatic, being a consequence of the gauge invariance of the elementary vertex $e(p_2 - p_1)_\mu$. Current algebra sum rules require [10] that be the residue of a fixed angular momentum pole at $J = 1$ in the crossed channel of the Compton amplitude for the scattering of a charged photon by a pion. We satisfy this by defining the compton amplitude by the Feynman graphs of Fig. 3. It can be shown easily that the residue of the fixed pole in question is obtainable from the graphs of Fig. 3 by "pinching" together points 1 and 2, thereby yielding the form factor.

Our plan is then to take the Feynman graphs of Fig. 2 and substitute for A a suitable dual amplitude. It should be noted that our approach is different from one in which one first tries to calculate approximately $\text{Im} F_\pi(t)$ and then use a dispersion relation to obtain $F_\pi(t)$. The latter approach would require a knowledge of the amplitude for $2\pi \to$ (arbitrary hadrons), and we want to avoid having to make models for that. In our approach the effect of the many hadron intermediate states are swept under A.

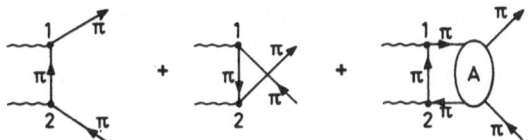

Fig. 3. Our model for pion Compton scattering

We decompose the pion amplitude A into the contributions:

$$A = A_{\text{Pom}} + A_\varrho,\tag{5}$$

where A_{Pom} contains the Pomeron trajectory and has no resonance poles, while A_ϱ is approximated by the Veneziano amplitude [11] which contains only resonance poles, the lowest one being the ϱ. Since the t-channel has isospin 1, we have [12]

$$A_\varrho(t, s) = -\beta [V(t, s) - V(t, u)],\tag{6}$$

$$V(t, s) = \Gamma(1 - \alpha_t)\,\Gamma(1 - \alpha_s)/\Gamma(1 - \alpha_t - \alpha_s),\tag{7}$$

$$\alpha_t = a + bt.\tag{8}$$

The trajectory α_t is completely determined by the conditions

$$\alpha(m_\varrho^2) = 1,$$
$$\alpha(m_\pi^2) = \tfrac{1}{2}.\tag{9}$$

The first states that the ϱ has spin 1, and the second is required to satisfy the Adler condition [12]. They lead to

$$a = \tfrac{1}{2} - \tfrac{1}{2}m_\pi^2(m_\varrho^2 - m_\pi^2)^{-1} \approx \tfrac{1}{2},$$
$$b = \tfrac{1}{2}(m_\varrho^2 - m_\pi^2)^{-1} \approx 1\,(\text{GeV}/c)^{-2}.\tag{10}$$

The constant β in (6) can be related to ϱ meson parameters by calculating the decay rate $\varrho \to 2\pi$ within the Veneziano model, leading to the relation

$$\beta = 16\pi(6\,\Gamma_\varrho/m_\varrho),\tag{11}$$

where Γ_ϱ is the ϱ decay width. Noting the experimental relation $6\Gamma_\varrho/m_\varrho \approx 1$, we obtain

$$\beta \approx 16\pi. \tag{12}$$

Thus there are no unknown parameters in A.

An important assumption is made here, namely that the amplitude A_ϱ is the same as that on the mass shell. The two intermediate pions, of course, go off the mass shell in the Feynman graph of Fig. 2. How the amplitude should depend on their masses is an open question. In field theories, such dependences are determined by the way current operators

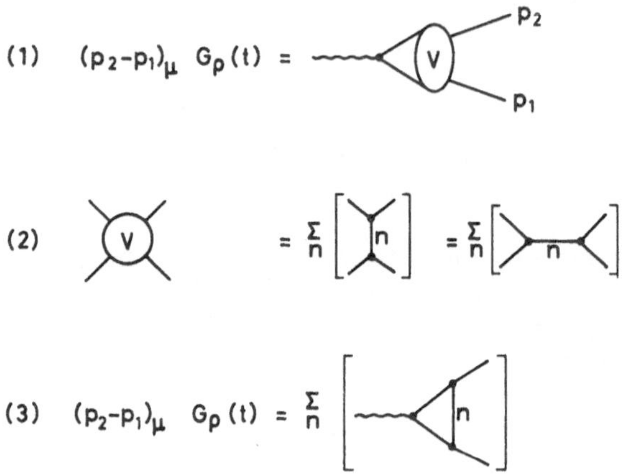

Fig. 4. Calculation of the form factor

are constructed out of field operators, in accordance with current conservation and assumptions of locality. We do not have such principles here. Therefore we make the simplest assumption, albeit a strong one, that there is no mass dependence. The consequence of this assumption will be clear, as will be the consequence of modifying it in other simple ways.

Corresponding to the decomposition (5), the form factor can be written as

$$F_\pi(t) = 1 + G_{\text{Pom}}(t) + G_\varrho(t), \tag{13}$$

and we shall calculate $G_\varrho(t)$. To sketch the calculation we refer to the graphs of Fig. 4, where on the first line $G_\varrho(t)$ is represented as an Feynman integral involving a Veneziano amplitude V. On the second line we note that the Veneziano amplitude is dual, that is, it can be represented as a sum of poles, either in the s or in the t channel. Thus, we see on line three that $G_\varrho(t)$ can be expressed as a sum of triangular graphs. Each triangular graph can be evaluated using standard Feynman graph techniques, and the answer turns out to be an integral whose integrand is of the form $(t + x_n)^{-1}$, where x_n is some complicated function independent of t. The important point is that it is again of the form of a pole in t. Summing over these pole terms again gives back a Veneziano amplitude. In this manner we obtain

$$G_\varrho(t) = (\beta/4\pi^2 t) \int_0^\infty dx \, V(t, -x) \, g(t, x), \tag{14}$$

where the function $g(t, x)$ is explicit given in Ref. [9]. We give here only the simple form it approaches when the pion mass is neglected:

$$g(t, x) \xrightarrow[m_\pi \to 0]{} (\tfrac{1}{2} - x/t) \ln(1 - t/x) - 1. \tag{15}$$

The logarithmic factor gives rise to a branch cut of $G_\varrho(t)$ from $t = 0$ to ∞ ($4m_\pi^2$ to ∞, if m_π had been kept finite).

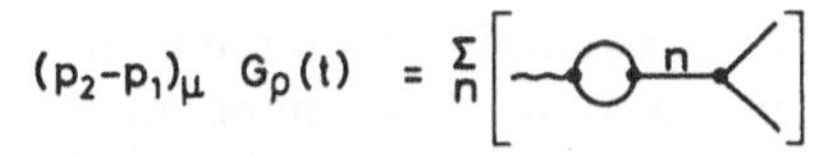

$$(p_2 - p_1)_\mu \ G_\rho(t) \ = \ \sum_n \left[\!-\!\!\bigcirc\!\!-^n\!\!\!< \right]$$

Fig. 5. Why second order poles occur

This elastic cut, of course, arises just from the Feynman propagator used for the two intermediate pions. Any inelastic cut that $G_\varrho(t)$ might have can arise only from the cuts of $V(t, -x)$, which in the Veneziano model are approximated by an infinite distribution of poles.

From (14) and (7), we see that $G_\varrho(t)$ is proportional to $\Gamma(1 - \alpha_t)$, which has simple poles at $\alpha_t = 1, 2, 3, \ldots$ corresponding to an infinite number of vector mesons $\varrho, \varrho', \varrho'', \ldots$, with equally spaced squared masses. When one attempts to calculate the residues of these poles from (14) by setting $\alpha_t = $ integer, one finds that they diverge. If one first calculates the residue function for $\alpha_t \neq n$, and then approaches $\alpha_t = n$, one finds that the residue function itself has a simple pole at $\alpha_t = n$. Therefore near $\alpha_t = n$, $G_\varrho(t)$ has a second order pole, plus a simple pole. This can be seen quite directly by noting that by duality we could have formally replaced line 3 of Fig. 4 by the graphs shown in Fig. 5. We see that each graph separately diverges because of the self-energy loop, and the divergence is directly traced to the assumption that A has no mass dependence when extrapolated off the mass shell.

It is not difficult to show that if the amplitude is damped by a factor $(m^2)^{-n}$ for each external mass m off the mass shell, then the first n vector mesons $\varrho, \varrho', \varrho'', \ldots$ will appear as simple poles, but the $(n + 1)$ st and higher one will continue to be second order poles. An exponential mass damping will make them all simple poles. Since there is no *a priori* reason why vector mesons should appear as simple poles in the form factor, we stick to the original version of the model, which is free of arbitrary parameters, and certainly more intriguing.

Let us examine $G_\varrho(t)$ in more detail near the vector meson poles, for in these neighborhoods $F_\pi(t) \approx G_\varrho(t)$. Thus, at the ϱ and ϱ' poles we have

$$\begin{aligned} F_\pi(t) &\approx (\beta/48\pi^2)\,(1-a)^{-1}\,b^{-2}\,(t-m_\varrho^2)^{-2} \\ &\quad + iO(t-m_\varrho^2)^{-1}, \qquad (t \approx m_\varrho^2), \end{aligned} \tag{16}$$

$$\begin{aligned} F_\pi(t) &\approx (\beta/48\pi^2)\,(2-a)\,(3a-1)^{-1}\,b^{-2}(t-m_\varrho'^2)^{-2} \\ &\quad + iO(t-m_\varrho'^2)^{-1}, \qquad (t \approx m_\varrho'^2), \end{aligned} \tag{17}$$

where $bm_\varrho^2 \approx \frac{1}{2}$, $bm_\varrho'^2 \approx \frac{3}{2}$. Note that the second order poles are real, while the simple poles are pure imaginary. For $a = \frac{1}{2}$ the residue of the ϱ' second order pole is $\frac{3}{2}$ that of the ϱ.

These poles, however, are on the real t-axis, because the Veneziano model is a zero-width approximation. How might these results be changed if widths are taken into account? We can, of course, only offer a guess, for we do not know the correct way to unitarize the Veneziano model. Since we know experimentally that $\varrho \to 2\pi$ is the overwhelmingly predominant decay mode of the ϱ, we may assume that near the ϱ pole it is a good approximation to assume elastic unitarity, which requires

$$\operatorname{Im} F_\pi(t) = F_\pi^*(t)\, T_\pi(t), \tag{18}$$

where $T_\pi(t)$ is the π-π p-wave, $I = 1$ scattering amplitude at squared energy t.

Elastic unitarity again requires

$$\operatorname{Im} T_\pi = T_\pi^*\, T_\pi. \tag{19}$$

Therefore the general solution to (18) is

$$F_\pi(t) = R(t)\, T_\pi(t), \tag{20}$$

where $R(t)$ is a real analytic function. The entire content of (20) is that the phase of $F_\pi(t)$ is the same as the phase of $T_\pi(t)$, i.e. the p-wave, $I = 1$ phase shift for π-π scattering. Let us assume that $T_\pi(t)$ contains the ϱ as a simple pole. There are then two possibilities: (a) The second order pole in $F_\pi(t)$ arises from a confluence of two poles of $T_\pi(t)$ in the limit of zero width; (b) It arises from a confluence of the ϱ pole in $T_\pi(t)$ with a pole in $R(t)$, in the limit of zero width. The first possibility can be ruled out as follows.

It can be shown that if T_π contains two simple poles close together, then (19) requires that their contribution to the phase shifts add, and this leads to

$$T_\pi \approx T_1 + T_2 + 2i\, T_1\, T_2\,, \tag{21}$$

$$T_n(t) = -\lambda_n(t - m_n^2 + i\lambda_n)^{-1}\,, \qquad (n = 1, 2)\,.$$

In the limit $m_1 \to m_2 \to m$, and $\lambda_1 \to \lambda_2 \to 0$, we would have, by (20),

$$F_\pi(t) \approx R(t)\left[-(\lambda_1 + \lambda_2)/(t - m^2) + 2i\lambda_1\lambda_2/(t - m^2)^2\right]\,, \tag{22}$$

which disagrees with (16), because the second order pole is pure imaginary instead of real. Therefore we can only have the second possibility, and since $R(t)$ is real analytic, its poles must occur in complex conjugate pairs. This leads to

$$F_\pi(t) \approx -\lambda_\varrho(t - m_\varrho^2 + i\lambda_\varrho)^{-1}\left[C + D((t - M^2 + i\sigma)^{-1} + (t - M^2 - i\sigma)^{-1})\right] \tag{23}$$

which approaches the form (16) when $m \to M$ and $\lambda_\varrho \to \sigma \to 0$. It is interesting that (23) is of similar form to (3), if m_ϱ and M are further apart than the widths.

To make some rough comparisons with experiments without having to adjust the numerous constants in (23), we may just take (16) and replace m_ϱ by $m_\varrho - i\Gamma_\varrho/2$, where $6\Gamma_\varrho/m_\varrho \approx 1$. We then obtain, near $t = m_\varrho^2$,

$$|F_\pi(t)|^2 \approx (6\pi b^2)^{-2}\left[(t - m_\varrho^2)^2 + (m_\varrho\Gamma_\varrho)^2\right]^{-2}\,. \tag{24}$$

In particular, at the ϱ peak

$$|F_\pi(m_\varrho^2)|^2 = (6\pi)^{-2}(b\,m_\varrho\Gamma_\varrho)^{-4} \approx 30\,, \tag{25}$$

which lies within experimental accuracy [2, 3]. If we simulate (24) with an equivalent simple Breit-Wigner formula of the same height and width, we obtain

$$|F_{\mathrm{eff}}(t)|^2 \approx (6\pi)^{-2}(m_\varrho\Gamma_{\mathrm{eff}})^2(b\,m_\varrho\Gamma_\varrho)^{-4}\left[(t - m_\varrho^2)^2 + (m_\varrho\Gamma_{\mathrm{eff}})^2\right]^{-1}$$
$$\Gamma_{\mathrm{eff}} = (2^{\frac{1}{2}} - 1)^{\frac{1}{2}}\Gamma_\varrho = 0.65\,\Gamma_\varrho\,, \tag{26}$$

which could explain the apparently smaller width of the ϱ in colliding beam experiments.

Other properties of $G_\varrho(t)$ would be less directly related to observations, because of the presence of the term $1 + G_{\mathrm{Pom}}(t)$. The contribution of $G_\varrho(t)$ to r_π, ignoring $G_{\mathrm{Pom}}(t)$, is divergent as $m_\pi \to 0$. The leading divergent term gives

$$[6\,G_\varrho'(0)]^{\frac{1}{2}} \xrightarrow[m_\pi \to 0]{} [12b\ln(1/b\,m_\pi^2)]^{\frac{1}{2}} = 1.4 \times 10^{-13}\ \mathrm{cm}\,, \tag{27}$$

which is of the same order of magnitude as the experimental value (1).

At $t = 0$ we find that $G_\varrho(t)$ vanishes in the limit $m_\pi \to 0$:

$$G_\varrho(0) \xrightarrow[m_\pi \to 0]{} -\beta b\,m_\pi^2/8\pi\,. \tag{28}$$

This fact is a direct consequence of the Adler condition that A_ϱ satisfies. Hence we expect $G_{\mathrm{Pom}}(0)$ to behave similarly. To order $b\,m_\pi^2$, therefore, the normalization condition $F_\pi(0) = 1$ is automatically satisfied. The cancellation $G_\varrho(0) + G_{\mathrm{Pom}}(0) = 0$ need be enforced only when we "turn on" the pion mass.

The asymptotic form of $F_\pi(t)$ is of particular interest, but it can only be a subject of conjecture for us. We do know by direct calculation from (14) that

$$G_\varrho(t) \xrightarrow[|t| \to \infty]{} - (\beta/8\pi^2)\, \Gamma(1 - a)\,(- bt)^{a-1}\,[1 + O(1/\ln bt)]\,. \qquad (29)$$

Assuming that $G_{\mathrm{Pom}}(t)$ behaves similarly except we set $a = 1$, ignoring the divergence of $\Gamma(1 - a)$, we would conclude that

$$G_{\mathrm{Pom}}(t) \xrightarrow[|t| \to \infty]{} \text{constant}\,, \qquad (30)$$

and it is very templing to conjecture that this constant should be exactly -1, thus cancelling the "contact term". This would lead to

$$F_\pi(t) \xrightarrow[|t| \to \infty]{} t^{-\frac{1}{2}}\,. \qquad (31)$$

It is difficult to construct a model similar to ours for the nucleon form factor, for the Veneziano amplitude involving fermion is unknown. However, if one writes the usual Veneziano amplitudes for the invariant amplitudes (thus ignoring difficulties with the occurence of parity doublet), one obtains in place of (31)

$$F_{\mathrm{Nucleon}}(t) \xrightarrow[|t| \to \infty]{} t^{\alpha_N(0) - \frac{3}{2}} = t^{-2} \qquad (32)$$

when $\alpha_N(0)$ is the intercept of the nucleon trajectory, taken to be $-\frac{1}{2}$.

References

1. Mistretta, C., *et al.*: Phys. Rev. Letters **20**, 1523 (1968).
2. Auslander, V. L., *et al.*: Phys. Letters **25** B, 433 (1967).
3. Augustin, J. E., *et al.*: Phys. Letters **28** B, 508 (1969).
4. Trigante, D., Wataghin, V.: (to be published); – Wataghin, V.: Nucl. Phys. B **10**, 109 (1969).
5. Lee, T. D., Wick, G. C.: Nucl. Phys. B **9**, 209 (1969).
6. Schwinger, J.: Phys. Rev. D **3**, 1967 (1971).
7. Kroll, N., Lee, T. D., Zumino, B.: Phys. Rev. **157**, 1376 (1967).
8. See Brower, R. C., Weis, J. H.: Phys. Rev. **188**, 2486, 2495 (1969) and references given there.
9. Gerstein, I. S., Gottfried, K., Huang, K.: Phys. Rev. Letters **24**, 294 (1970).
10. Bronzan, J. B., Gerstein, I. S., Lee, B., Low, F. E.: Phys. Rev. **157**, 1448 (1967).
11. Veneziano, G.: Nuovo Cimento **57** A, 190 (1968).
12. Lovelace, C.: Phys. Letters **28** B, 265 (1968).

Dr. Kerson Huang
Laboratory for Nuclear Science and Physics Department
Massachusetts Institute of Technology
Cambridge, Mass., USA

Deep Inelastic Hadronic Scattering in Dual-Resonance Model

KERSON HUANG

There has been much interest recently in the so-called inclusive hadronic reactions, namely a reaction of the type

$$A + B \rightarrow X + \text{anything} \tag{1}$$

where A, B, and X are hadrons. Most of the works in the literature (1) have dealt with special regions near the phase space boundaries of the reaction, in which the general behavior of the cross section can be predicted by a new type of Regge phenomenology, without having to assume a detailed dynamical model. In this respect they are similar to the Regge phenomenology of two-body near-forward scattering.

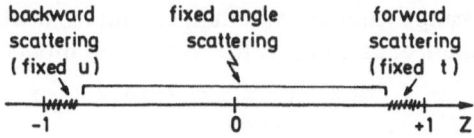

Fig. 1. Phase space for two-body reaction at given s

I would like to explore the behavior of such inclusive reactions in the phase space far from the boundaries – a region that I shall refer to as the "deep inelastic region" in analogy to the well-known case of electron scattering. Here one can make predictions only if one makes detailed dynamical assumptions. In this respect it is similar to wide-angle scattering in two-body processes. Before we go into the subject, it is helpful, for orientation, to review those aspects of two-body processes that are relevant to our developments.

For definiteness let us consider $p - p$ elastic scattering at high energies. The kinematic is completely specified by two numbers: the squared CM energy s, and the squared 4-momentum transfer $t = -2p^2 \cos\theta$, where p is the magnitude of the 4-momentum, and θ the scattering angle, all in the CM frame. For fixed s, therefore, the phase space θ may be represented by the line segment $-1 \leq z \leq +1$, where $z = \cos\theta$, as shown in Fig. 1. The shaded positions near the phase space boundaries are those amenable to conventional phenomenological Regge analysis, with the differential cross section $d\sigma/d\Omega$ behaving as $\beta(t) \, s^{a(t)}$, or $\beta(u) \, s^{\alpha(u)}$, respectively. The extensions of these shaded portions shrink to zero as $s \to \infty$. In the wide-angle region, which is almost all of phase space as $s \to \infty$, we need dynamical models, of which there are few, and none so far is completely satisfactory. The experimental data in the $10 - 30 \, \text{GeV}/c$ range of incident laboratory momentum seem to show a remarkably simple scaling behavior. Orear [2] suggested the empirical formula

$$\sigma_{el}(s, z) \equiv d\sigma/d\Omega \approx \text{Const. } e^{-a p \sin\theta}, \tag{2}$$

where a is approximately independent of the energy, and has a value of about $(160 \, \text{MeV}/c)^{-1}$. Krisch [3] suggested that a more accurate fit can be obtained by replacing $p\sin\theta$ by $(p\sin\theta)^2$. On the basis of a high statistics experiment, the CERN group [4] found that a still better empirical formula is

$$\sigma_{el}(s, z) \approx \text{Const. } e^{-A s \sin\theta}. \tag{3}$$

None of these really fit the data in all the fine structure, but they reproduce the gross features rather well.

On the theoretical side, there is only one dynamical model that is reasonably precise, based on premises subject to other independent tests such as crossing symmetry and duality, and from which predictions can be easily obtained, and that is the Veneziano model [5]. It predicts

$$\sigma_{el}(s, z) \xrightarrow[z \text{ fixed}]{s \to \infty} C \, e^{b s g(z)},$$
$$g(z) = (1 + z) \ln\tfrac{1}{2}(1 + z) + (1 - z) \ln\tfrac{1}{2}(1 - z), \tag{4}$$

where the quantity C generally depends on s and z, and also depends on specific choices of terms in the Veneziano model (e.g. st, su, or tu terms); but for large s, it varies at most like a power of s, and can be regarded as a constant compared to the exponential factor. A very good numerical approximation to the function $g(z)$, for the whole range of $z = \cos\theta$, is

$$g(z) \approx -(2\ln 2) \, (\sin\theta)^{\frac{3}{2}}. \tag{5}$$

Thus by comparing (4) and (5) to (2) and (3), we see that the Veneziano model gives a prediction that fits well the gross features of experimental data, provided we regard b as an adjustable parameter.

Now within the Veneziano model b is fixed, being the universal slope of nonvacuum Regge trajectories, and should have a value $b \approx 1 \, (\text{GeV}/c)^{-2}$. On the other hand to fit the data one needs $b \approx \frac{1}{2} \, (\text{GeV}/c)^{-2}$. One might "explain" this by saying that the Veneziano model leaves out the Pomeranchuk trajectory, which is expected to have a smaller slope, and that if one known how to incorporate it into the Veneziano model, one might get (4) with a smaller effective value of b. We therefore do not really have a prediction from a theory, but merely a suggestive result from an incomplete theory, the improvement of which, of course, has been the major concern of many theorists.

In the same spirit as the application of the Veneziano model to elastic widge-angle scattering, let us apply the dual-resonance model – the generalized N-point Veneziano amplitude – to the inclusive hadronic reaction (1). The kinematic is now specified by two numbers, apart from the squared CM energy s. Many experimentalists use $q_\perp$ and $q_\parallel$,

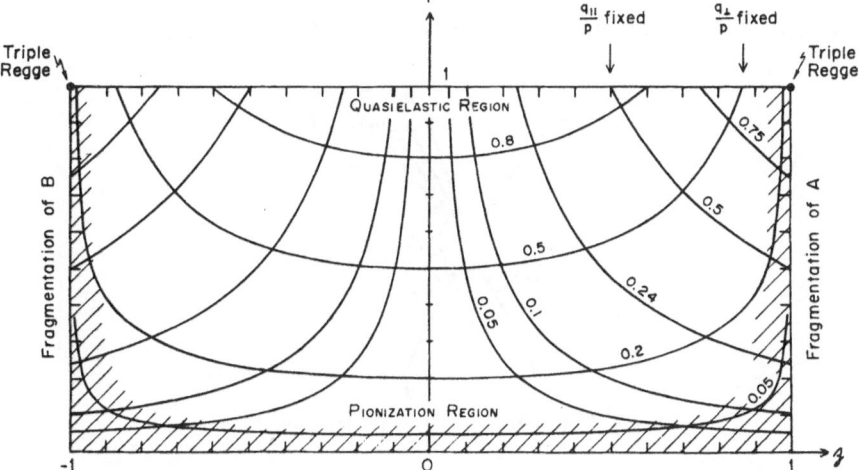

Fig. 2. Phase space for inclusive reaction at given s. Deep inelastic region is entire rectangle except for the shaded strip, whose width decreases as s increases. Lines of constant $q_\parallel/p = rz$, constant $q_\perp/p = r(1-z^2)^{\frac{1}{2}}$ are also shown

the transverse and longitudinal momentum of the produced particle X in the CM frame. As we shall see, however, it is more convenient to use the CM momentum q of the produced particle, and its CM angle θ with respect to the incident direction. We shall introduce

$$r = q/p, \quad (0 \le r \le 1)$$
$$z = \cos\theta, \quad (-1 \le z \le 1), \tag{6}$$

where p is the incident momentum in the CM frame. Thus, for given s, the two-dimensional phase space is represented by the $r - z$ plane in Fig. 2, on which is also shown lines of constant $q_\perp/p$ and $q_\parallel/p$. The shaded strip near the phase space boundaries $|z| = 1$ and $r = 0$ are regions analogous to the forward and backward scattering regions in two-body processes, in that the behavior of the cross section can be deduced from the new Regge analysis of A. H. Mueller (Ref. [1]). Other conventionally used kinematical variables are illustrated in Fig. 3, with definitions and asymptotic forms in terms of r and z given by

$$s \equiv (p_A + p_B)^2 \rightarrow 4p^2 \,,$$
$$t \equiv (p_A - p_X)^2 \rightarrow -\tfrac{1}{2}sr(1-z) \,,$$
$$u \equiv (p_B - p_X)^2 \rightarrow -\tfrac{1}{2}sr(1+z) \,,$$
$$M^2 \equiv (p_A + p_B - p_X)^2 = s + t + u - m_X^2 - m_A^2 - m_B^2$$
$$\rightarrow s(1-r) \,,$$

$$(7)$$

where "$\rightarrow$" denotes the limit $s \rightarrow \infty$ with r and z held fixed.

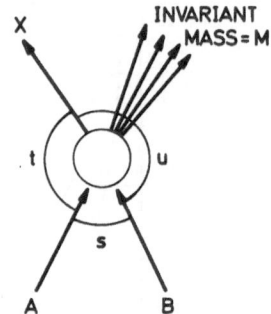

Fig. 3. Kinematical variables for inclusive reaction

The region near $z = 1$ is called the fragmentation region of the incident particle A, because as $s \rightarrow \infty$, the produced particle X carries off a finite fraction r of the incident momentum, with however a finite momentum transfer t from the incident particle. Thus it came off as a "fragment" of A. The differential cross section in this region is given by

$$d^2\sigma/dq\,d\Omega \approx [(1+r)/(1-r)]^{\alpha(t)} \,, \quad \text{(fragmentation region)} \quad (8)$$

where $\alpha(t)$ is an appropriate Regge trajectory. Similarly the region near $z = -1$ is the fragmentation region of the target B. The region near $r = 0$ is called the pionization region, from a terminology originating in cosmic-ray experiments, because even as $s \rightarrow \infty$, the produced particle X has finite momentum q. Mueller's analysis predicts the cross section to be a function of $q_\perp = q \sin\theta$ alone, but is unable to determine the functional form. DeTar *et al.* (Ref. [1]) calculated it explicitly from the dual-resonance model:

$$d^2\sigma/dq\,d\Omega \approx \exp(-4bq^2\sin^2\theta) \,, \quad \text{(pionization region)} \quad (9)$$

where b is the universal slope of linear Regge trajectories in that model. This yields a transverse momentum cut off of 500 MeV/c, which is in qualitative agreement with the experimental value of $300 - 500$ MeV/c.

We are interested in the region complement to the regions above (i.e. the region outside of the shaded strip in Fig. 2), and we shall call it the deep inelastic region, in analogy with ep inelastic scattering. For

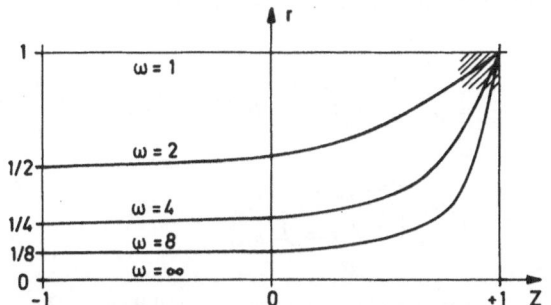

Fig. 4. Lines of constant $\omega = -2m_B\nu/t$ for inclusive reaction

comparison we give the conventionally used auxiliary variables in the latter case:

$$q^2 \equiv (p_X - p_A)^2 = t,$$
$$\nu \equiv p_B \cdot (p_A - p_X)/m_B = (t - u + m_B^2 - m_A^2)/2m_B, \tag{10}$$
$$\omega \equiv -2m_B\nu/q^2.$$

The deep inelastic region in ep scattering is defined as that in which $\nu \to \infty$ at fixed ω:

$$\omega \approx 1 - M^2/t \approx 1 + 2(1 - r)/r(1 - z). \tag{11}$$

On our $r - z$ plot, therefore, the lines of constant ω are as indicated in Fig. 4.

The deep inelastic region here is mathematically defined by the rule that we take the limit $s \to \infty$ first, with $r \neq 0, 1$ and $|z| \neq 1$ fixed. Only after this is done shall we approach the phase space boundaries. In particular the quasielastic region $r = 1$ means $s \to \infty, r \to 1$, in that order, so that even there the final state of the reaction contains many particles. As $s \to \infty$, the inelastic regions cover almost all of phase space; but of course the cross section is vanishingly small, most of it being concentrated in the fragmentation and pionization regions.

To calculate the inclusive cross section $d^2\sigma/dqd\Omega$ from the dual-resonance model, we proceed as did Mueller and DeTar *et al.* by first making use of a generalized optical theorem, which relates $d^2\sigma/dqd\Omega$

to the absorptive part of a 6-point function in the variable M^2:

$$d^2\sigma/dq\,d\Omega = \operatorname{Im} B_6(M^2),\tag{12}$$

which may be illustrated by the graphs in Fig. 5.

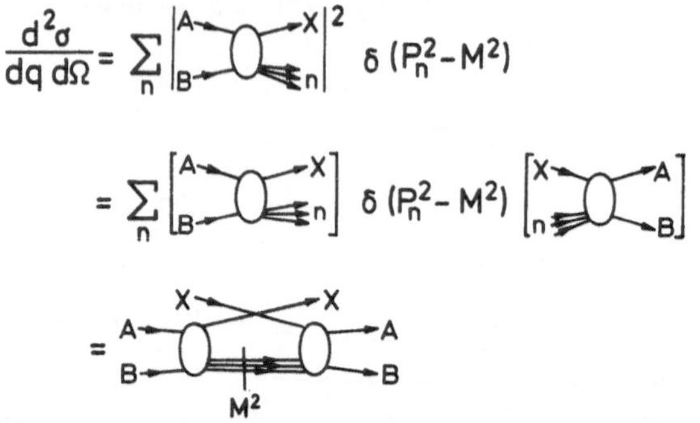

Fig. 5. Illustration of the generalized optical theorem [Eq. (12)]

The dual-resonance model is then used to calculate $\operatorname{Im} B_6$. Thus (12) is an exact statement implied by unitarity, but we approximate the right hand side by using the dual-resonance model, which does not satisfy unitarity. The dual-resonance model for B_6 is of the form

$$B_6(M^2) = \pi^{-1} \sum_{n=1}^{\infty} R(n)/(n - M^2),\tag{13}$$

where dependence on variables other than M^2 has been suppressed.

Therefore, strictly speaking

$$\operatorname{Im} B_6(M^2) = \sum_{n=1}^{\infty} R(n)\,\delta(n - M^2).\tag{14}$$

For $M^2 \to \infty$, however, we may perform a local average over M^2, and obtain

$$\operatorname{Im} B_6(M^2) \approx R(M^2).\tag{15}$$

This residue function is easily calculated, and has been given by DeTar *et al.* (Ref. [1]). We only have to take their result and evaluate it explicitly in the deep inelastic region. The result [6] turns out to be extremely simple and suggestive:

$$d^2\sigma/dq\,d\Omega \xrightarrow[\substack{s \to \infty \\ r, z \text{ fixed}}]{} \sigma_0 e^{bsG(r,z)},\tag{16}$$

$$G(r, z) = (1 + rz)\ln\tfrac{1}{2}(1 + rz) + (1 - rz)\ln\tfrac{1}{2}(1 - rz)$$
$$- (1 + r)\ln\tfrac{1}{2}(1 + r) - (1 - r)\ln\tfrac{1}{2}(1 - r),\tag{17}$$

where b is the universal slope of linear Regge trajectories in the model:

$$\alpha(t) = a + bt \,, \tag{18}$$

and σ_0 is a function of s, r, z, that varies at most like a power of s as $s \to \infty$. Its explicit form depends on the details of the reactions (i.e. quantum numbers of the participants). The function $G(r, z)$, however, is universal. For sufficient large s, σ_0 becomes unimportant, for according to (16)

$$F \equiv s^{-1} \, d^2\sigma/dq \, d\Omega \to bG(r, z) + O(\ln s/s) \,. \tag{19}$$

For comparison with existing data, however, we take σ_0 to be a constant:

$$\sigma_0 = 1 \text{ mb}/(\text{Sr} - \text{GeV}/c) \,. \tag{20}$$

a value consistent with the observed $p - p$ total cross section. By comparing (17) with (4) we obtain the interesting formula

$$\sigma_0^{-1} d^2\sigma/dq \, d\Omega \xrightarrow[\substack{s \to \infty \\ r,z \text{ fixed}}]{} \sigma_{\text{el}}(s, rz)/\sigma_{\text{el}}(s, r) \,. \tag{21}$$

That is, in the dual-resonance model, the inclusive cross section in the deep inelastic region is determined by the cross section for wide-angle elastic scattering.

The quasi-elastic limit $r \to 1$ is particularly interesting, for (21) gives

$$d^2\sigma/dq \, d\Omega \xrightarrow[\substack{s \to \infty \\ r \to 1}]{} \text{Const. } \sigma_{\text{el}}(r, z) \,. \tag{22}$$

We note from (7) that in this limit $t \gg M^2$, although $M^2 \to \infty$. Thus even though many particles are produced, the cross section is like that for elastic scattering, as long as the momentum transfer is much larger than M^2.

We emphasize that (16) and (17) apply only to the deep inelastic region. They do, however, approach limits that are qualitatively correct when one approaches the phase space boundaries $|z| = 1$ or $r = 0$. For example, as $r \to 0$, one recovers the pionization limit:

$$\ln(d^2\sigma/dq \, d\Omega) \to -4bq^2 \sin^2\theta \,. \tag{23}$$

As $|z| \to 1$, one recovers the fragmentation limit:

$$\ln(d^2\sigma/dq \, d\Omega) \to bt \ln[(1+r)/(1-r)] \,, \tag{24}$$

which differs from the correct expression (8) only in that bt should have been $a + bt$, where a is the Regge intercept.

We now compare our model with data by examining the ratio

$$b \equiv s^{-1} \ln(d^2\sigma/dq \, d\Omega)_{\text{CM}}/G(r, z) \,, \tag{25}$$

when $(d^2\sigma/dq\,d\Omega)_{CM}$ is the experimental cross section in the CM frame, in mb/(Sr–GeV/c). Scaling is tested by seeing whether b is independent of s. The dual-resonance form (17) of $G(r,z)$ is tested by the additional requirement that b be a constant, whose value would then be the slope parameter of some effective Regge trajectory.

We use the data at 30 and 20 GeV/c of Anderson *et al.* [7] and those at 12.5 GeV/c of Akerlof *et al.* [8]. To exclude the "strip", we keep only values of $|z| < 0.7$, in all the data. The results are shown in Figs. 6–9, which are self-explanatory. The extensive data at 19.2 GeV/c of Allaby *et al.* [9], unfortunately fall entirely in the fragmentation and pionization regions, and hence cannot be used for our purpose.

The validity of scaling, and the specific form of $G(r,z)$, seem to be borne out. The general feature is that $s^{-1}\ln(d^2\sigma/dq\,d\Omega)$ approaches the universal function $G(r,z)$ as $s \to \infty, z \to 0, r \to 1$. For fixed r, the rate of approach is faster the larger the s, and the smaller the z. This is illustrated most clearly in the π^+ data in Fig. 7. As an example of the quality of the fit, we compare two data points in the 30 GeV/c proton spectra:

r	z	$G(r,z)$	b
0.17	-0.24	-0.029	0.32
0.57	0.36	0.3	0.33

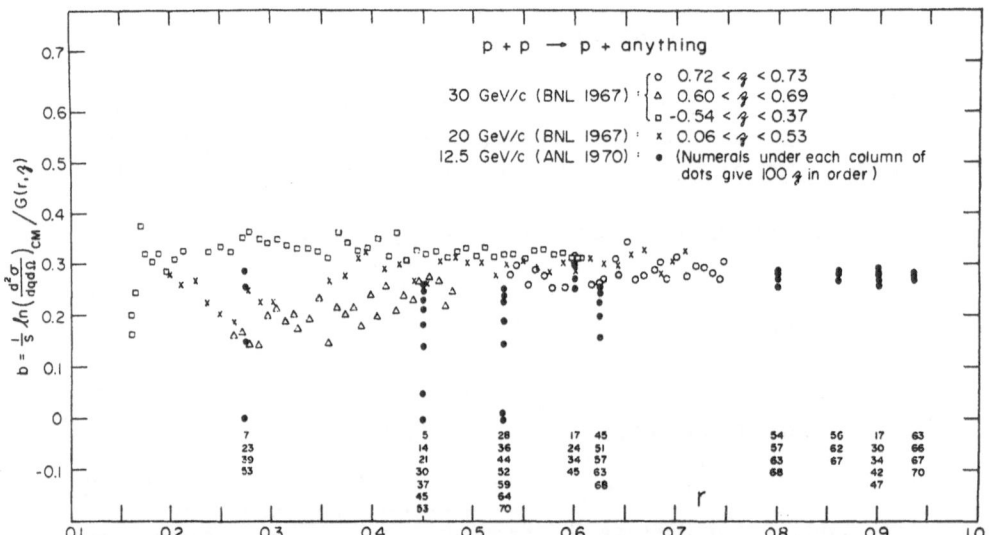

Figs. 6–9. Plots of experimental $s^{-1}\ln(d^2\sigma/dq\,d\Omega)_{CM}$ divided by known function $G(r,z)$, versus r for different values of z. The experimental differential cross section $(d^2\sigma/dq\,d\Omega)_{CM}$ is in units of mb/(Sr – GeV/c), and s is in (GeV)^2. Data at 30 and 20 GeV/c (BNL 1967) are taken from Ref. [7], and those at 12.5 GeV/c (ANL 1970) are taken from Ref. [8].

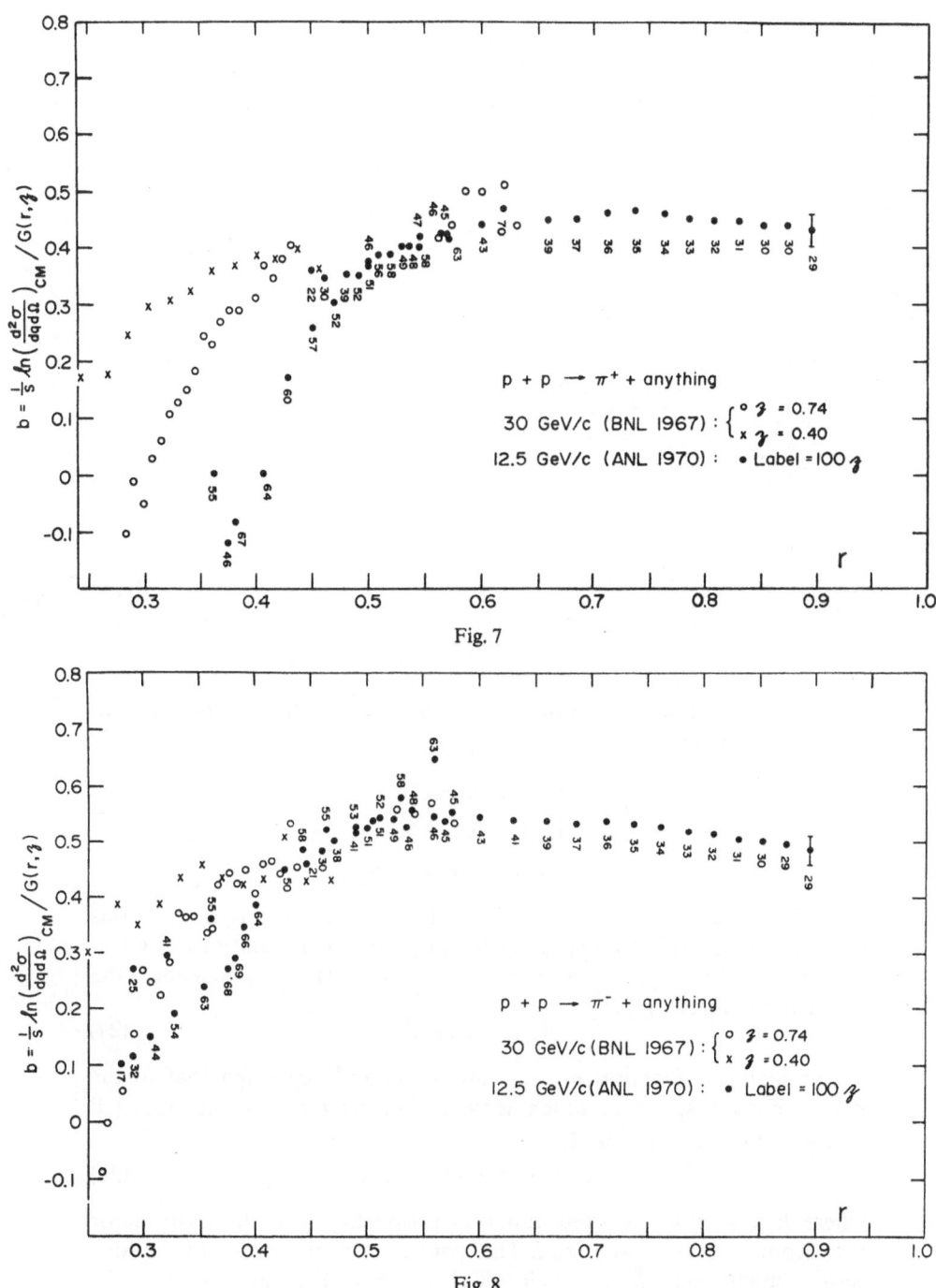

Fig. 7

Fig. 8

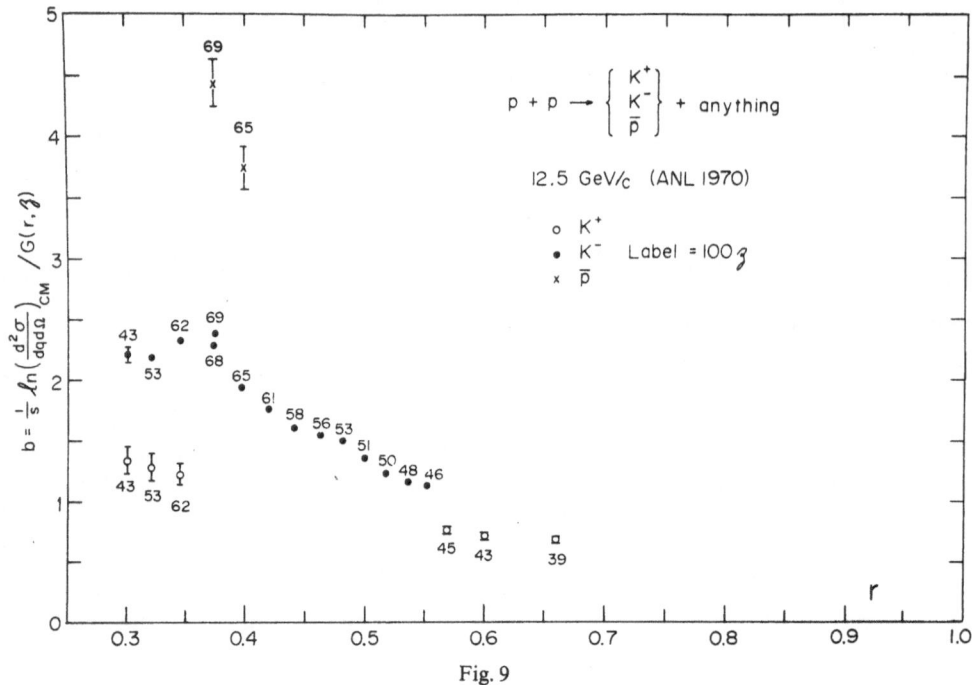

Fig. 9

The effective slope parameter b seems to be different for different outgoing particles:

$$b \approx \begin{cases} 0.3 \text{ for } X = p \\ 0.4 \text{ for } X = \pi^+ \\ 0.5 \text{ for } X = \pi^- \\ 0.5 \text{ for } X = \tilde{K}^+, K^- . \end{cases} \qquad (26)$$

The relevance of these results to the electromagnetic interactions of hadronics lies in a comparison with some recent speculations by Chen and Harte [12] and by Berman and Jacob [13]. They extended the Wu-Yang conjecture [14]

$$\sigma_{ep \to ep} \approx (\sigma_{el})^{\frac{1}{2}} \qquad (27)$$

to the inclusive reaction $pp \to p +$ anything, and suggested that in the region of phase space for which M^2/t remains fixed as $r \to 1$, as indicated by the shaded area in Fig. 4,

$$d^2\sigma/dq\,d\Omega \approx (\sigma_{el})^{\frac{1}{2}} R , \qquad (28)$$

where R is a slowly varying function proportional to the form factor νW_2 from $e + p \to e +$ anything. This conjecture is not borne out by our result, which resembles more closely (27) with $(\sigma_{el})^{\frac{1}{2}}$ replaced by σ_{el}.

References

1. Benecke, J., Chou, T. T., Yang, C. N., Yen, E.: Phys. Rev. **188**, 2153 (1968); — Feynman, R. P.: Phys. Rev. Letters **23**, 1415 (1969); — Mueller, A.: Phys. Rev. D**2**, 2363 (1970; — DeTar, C., Kang, K., Tan, C. I., Weis, J.: MIT Preprint (1971), to be published in Phys. Rev. See also Cheng, H., Wu, T. T.: Phys. Rev. Letters **23**, 1311 (1969).
2. Orear, J.: Phys. Rev. Letters **12**, 112 (1964).
3. Krisch, A. D.: Phys. Rev. Letters **11**, 217 (1963).
4. Allaby, J. V., *et al.*: Phys. Letters **25**B, 156 (1967).
5. Veneziano, G.: Nuovo Cimento, **57**A, 190 (1968).
6. Huang, K., Segre, G.: MIT preprint (1971).
7. Anderson, E. W., *et al.*: Phys. Rev. Letters **19**, 198 (1967).
8. Akerlof, C. W., *et al.*: Phys. Rev. D**3**, 645 (1971).
9. Allaby, J. V., *et al.*: CERN NP preprint 70–12 (1970).
10. Chen, M. S., Harte, J.: Phys. Rev. D**1**, 2603 (1970).
11. Berman, S., Jacob, M.: Phys. Rev. Letters **25**, 1683 (1970).
12. Wu, T. T., Yang, C. N.: Phys. Rev. **137**, B708 (1965).

Dr. Kerson Huang
Laboratory for Nuclear Science and Department of Physics
Massachusetts Institute of Technology
Cambridge, Mass., USA

Low Energy Theorems and Photo- and Electroproduction Near Threshold by Current Algebra

G. Furlan, N. Paver, and C. Verzegnassi

Contents

I. Introduction

These lectures are devoted to a simple current algebra description of low energy pion physics. The motivation for a renewed interest in the subject is represented by the fact that there is some feeling, at present, to consider the small pion mass, rather than a dynamical accident, as the strong hint for an approximate $SU(2) \times SU(2)$ symmetry of the hadron world. In order to establish a common framework let us quickly remind a few known facts about currents and current algebra.

Roughly speaking, one can say that current algebra can be considered as the successful attempt to generalize some outstanding results of electrodynamics, like the non-renormalization of the electric charge by strong interactions and the low energy photon theorems, to the other currents used to describe some important properties of hadrons. This is achieved by applying to j_μ^{em} a $SU(3)$ rotation and a parity transformation, i.e. a $SU(3) \times SU(3)$ transformation. As a consequence we have to deal with two octets of currents, vector and axial vector, V_μ^α and A_μ^α, $\alpha = 1 \ldots 8$.

Some of them are used to describe the electromagnetic interactions (V_μ^3, V_μ^8) and the weak interactions $(V_\mu^{1,2,4,5}, A_\mu^{1,2,4,5})$ between hadrons and leptons.

The following properties of these currents represent important achievements in our understanding of elementary particle physics.

a) Algebraic Properties

Consider the weak charges associated with the currents

$$Q^\alpha = \int dx\, V_0^\alpha(x), \qquad \bar{Q}^\alpha = \int dx\, A_0^\alpha(x). \tag{1.1}$$

These objects are assumed to be the generators of the $SU(3) \times SU(3)$ algebra and to obey therefore the set of equal time commutation relations

$$[Q^\alpha, Q^\beta] = i f_{\alpha\beta\gamma} Q^\gamma$$
$$[Q^\alpha, \bar{Q}^\beta] = i f_{\alpha\beta\gamma} \bar{Q}^\gamma \tag{1.2}$$
$$[\bar{Q}^\alpha, \bar{Q}^\beta] = i f_{\alpha\beta\gamma} Q^\gamma.$$

The proposed identification represents a non trivial bridge between the symmetry properties of the hadron world (i.e. behaviour under $SU(3) \times SU(3)$) and the quantities used to describe the lowest order weak and electromagnetic interactions of hadrons. Furthermore the validity of the above commutation relations is absolute, independent on whether the charges are conserved or not.

b) Conservation Properties

The Vector Case

We have

$$\partial^\mu V_\mu^{3,8} = 0,$$

and

$$\partial^\mu V_\mu^{1,2} = 0$$

only in the $SU(2)$ limit, i.e. in the absence of electromagnetism (and weak interactions). Moreover

$$\partial^\mu V_\mu^{4\cdots7} = 0$$

only in the $SU(3)$ limit, i.e. in the absence of "medium strong interactions".

These conservation properties reflect well known *algebraic symmetries* (or normal symmetries). One then obtains, in the symmetry limit, familiar results like the classification of hadrons in degenerate multiplets corresponding to $SU(3)$ irreducible representations, non-renormalization theorems for all charges, relations among vertex functions and so on. This realization of $SU(3)$ seems to be the most adherent to the real state of affairs. Thus, even if some $SU(3)$ charges are no longer conserved in nature and corrections to the symmetry results appear, one can still say that there are evident traces, in the physical world, of the underlying symmetry. Quantitatively this means that if one writes the hadron Hamiltonian as

$$H = H_0 + \varepsilon H_1 , \tag{1.3}$$

where H_0 is $SU(3)$ invariant while H_1 is not, the breaking parameter ε is small enough to justify attention to the zeroth and first order predictions.

The Axial Vector Case

Let us concentrate on the isospin axial currents and charges A_μ^α, $\bar{Q}^\alpha$, $\alpha = 1, 2, 3$. The first hint about the peculiar nature of the conservation limit ("chiral" limit) comes from the existence of a longitudinal component of the current, associated with a 0^- particle, the pion:

$$\langle 0 | A_\mu^\alpha | \pi^\beta \rangle = i \delta_{\alpha\beta} \pi_\mu f_\pi \tag{1.4}$$

and

$$\langle 0 | \partial^\mu A_\mu^\alpha | \pi^\beta \rangle = m_\pi^2 f_\pi \delta_{\alpha\beta} ,$$

showing that the symmetry limit is realized either by $m_\pi = 0$ or $f_\pi = 0$, namely either the particle is massless or it is completely decoupled from the current. While the second alternative is easily recognized to correspond to normal symmetries, the peculiar realization by massless and spinless bosons characterizes the so-called "spontaneously broken symmetries", à la Goldstone. A further argument comes from considering the nucleon matrix elements

$$\langle p_2 | A_\mu^\alpha | p_1 \rangle = \bar{u}_2 (\tau_\alpha/2) [\gamma_5 \gamma_\mu G_A(\Delta^2) + \Delta_\mu \gamma_5 G_P(\Delta^2)] u_1 , \tag{1.6}$$

$$\Delta = p_2 - p_1$$

and

$$\langle p_2 | \partial^\mu A_\mu^\alpha | p_1 \rangle = i \bar{u}_2 \gamma_5 (\tau^\alpha/2) u_1 [-2 M G_A(\Delta^2) + \Delta^2 G_P(\Delta^2)] . \tag{1.7}$$

The conservation requirement taken at $\Delta^2 = 0$ shows that if

$$\Delta^2 G_P(\Delta^2)|_{\Delta^2 = 0} = 0$$

(as the case when $m_\pi \neq 0$) then

$$M G_A(0) = 0 \tag{1.8}$$

and one is left with consequences of the symmetry, $M = 0$ or $G_A(0) = 0$, which are both badly violated in Nature. A more appealing result comes from the assumption that the axial current is still coupled to a massless pion. In this case $G_P(\varDelta^2) \xrightarrow{\varDelta^2 \to 0} -2f_\pi g_{\pi N}/\varDelta^2$ and one finds the Goldberger-Treiman relation

$$g_{\pi N} = -M G_A(0)/f_\pi . \tag{1.9}$$

We can consider it as the first example of low energy theorem, expressing the nucleon-massless pion vertex in terms of weak interaction quantities. The agreement between the above theoretical prediction and the experimental value is quite reasonable[1] (within a 10 %). This and other successful predictions we shall discuss later, support the idea that there is an underlying $SU(2) \times SU(2)$ symmetry in Nature, corresponding to the conservation of isospin vector and axial vector currents and realized by $SU(2)$ multiplets and massless pions[2]. The immediate consequences of chiral symmetry are then of a dynamical nature, since one can derive exact low energy theorems for massless pion amplitudes rather than algebraic results.

Actually the pion has a small, but non vanishing, mass (on the hadron scale) and chiral symmetry is broken. However, we can say that axial currents still show a sort of "partial conservation", which is concretely realized if the divergences $\bar{D}^\alpha = \partial^\mu A_\mu^\alpha$ (and their matrix elements) are proportional to some positive power of the pion mass. This is the basic idea of the so-called PCAC hypothesis, which, more concretely, reads $\langle b| \bar{D}^\alpha |a\rangle \sim O(m_\pi^2)$. Again a unified way of expressing all this is to split the total Hamiltonian

$$H = H_0 + \varepsilon H_1 \tag{1.10}$$

in a part, H_0, $SU(2) \times SU(2)$ invariant plus a part, εH_1, responsible of chiral breaking:

$$\dot{\bar{Q}}^\alpha = \int \bar{D}^\alpha \, \mathrm{d}x = i[H, \bar{Q}^\alpha] = i\varepsilon[H_1, \bar{Q}^\alpha] . \tag{1.11}$$

Then ε is consistently determined to be proportional to m_π^2.

This scheme supplies the most suited frame to fit all soft pion theorems, which can thus be logically understood as a product of the current algebra commutators and of PCAC (i.e. approximate $SU(2) \times SU(2)$ symmetry). Since the experimental verification of these theorems is at hand, the problem of evaluating the "corrections" due to the finite pion mass is of some importance, both from the practical point of view of having predictions for physical pions, and with the aim of a better insight into the nature of the breaking.

We shall illustrate in these lectures a simple formalism, based on current commutators, to obtain a representation of physical pion ampli-

---

[1] With $f_\pi|_{exp} = 0.68 \, m_\pi$, $G_A(0)|_{exp} = -1.23$, one gets $g_{\pi N}|_{G.T.} \simeq 12.3$ while $g_{\pi N}|_{exp} \simeq 13.5$.

[2] It has to be noticed that one could look for normal realizations of chiral symmetry, resulting in degenerate parity multiplets or vanishing mass nucleons, but the Goldstone mechanism seems to be the preferred one.

tudes, where both the soft pion limit and the corrections are explicitly exhibited. Owing to the chosen realization of the symmetry, the latter will vanish with the pion mass and will thus depend both on matrix elements and on "higher" commutators involving the axial current divergence. The application of the approach will be limited to pion-nucleon scattering and pion-electroproduction (with special emphasis on the last reaction), namely to "old fashioned" pion physics, where the present low-energy description can be considered complementary to dispersion theory or to the static model. Although other phenomena, like K_{l_3} decay, can be treated with the same techniques, we prefer, for time reasons, to concentrate on reactions where there is a sharp distinction between the pion and the target and no $SU(3)$ considerations are required.

Thus we shall only occasionally mention the recent and interesting ideas about $SU(3) \times SU(3)$ as an approximate symmetry of strong interactions. This represents a natural generalization of the previous considerations, which arises combining $SU(3)$ and $SU(2) \times SU(2)$. The interplay between these two partial symmetries, the role of kaon PCAC, the $SU(3) \times SU(3)$ transformation properties of H_1 in Eq. (1.10), represent open and fascinating problems and we hope that the techniques we present here can represent a tool for a better understanding of those points.

In Section II we shall review the main features of the formalism and treat in some detail the pion-nucleon scattering. Sections III, IV will be devoted to photo- and electroproduction of one pion and to a comparison with experiment.

II. Low Energy Theorems

a) A Simple Formalism for Low Energy Pion Physics

One of the remarkable features of current algebra is that several techniques are available, rather than a single one, to exploit the fundamental commutation relations. The various approaches (chiral lagrangians, Ward identities, sum rules, etc.) have their own peculiar aspects but, after careful translation, the results they give are seen to be more or less equivalent. We choose to describe a simple technique of deriving low energy pion theorems, based upon the saturation of equal-time commutators. Of course, the commutators must involve the axial charges $\bar{Q}^\alpha$ ($\alpha = 1, 2, 3$) since the symmetry pattern has to be explicit.

Let us define the operators [1]

$$\begin{aligned}
\bar{Q}^\alpha_{L,R} &= \bar{Q}^\alpha \pm (i/m_\pi)\, \dot{\bar{Q}}^\alpha \\
\bar{Q}^\alpha_0 &= \bar{Q}^\alpha + (1/m_\pi^2)\, \ddot{\bar{Q}}^\alpha = \bar{Q}^\alpha_{L,R} \mp (i/m_\pi)\, \dot{\bar{Q}}^\alpha_{L,R}
\end{aligned} \qquad (2.1)$$

such that

$$\langle 0|\bar{Q}_L^\alpha|\pi^\beta\rangle = -\langle\pi^\beta|\bar{Q}_R^\alpha|0\rangle = 2\,i f_\pi m_\pi (2\pi)^3\,\delta_{\alpha\beta}\,\delta(\pi)$$
$$\langle\pi|\bar{Q}_L^\alpha|0\rangle = \langle 0|\bar{Q}_R^\alpha|\pi\rangle = \langle\pi|\bar{Q}_0^\alpha|0\rangle = \langle 0|\bar{Q}_0^\alpha|\pi\rangle = 0\,. \tag{2.2}$$

Making use of the relations

$$\dot{\bar{Q}}^\alpha = \int \mathrm{d}\boldsymbol{x}\,\bar{D}^\alpha(x)\,, \qquad \bar{D}^\alpha(x) = \partial^\mu A_\mu^\alpha(x)$$
$$\dot{\bar{Q}}_0^\alpha = m_\pi^{-2}\int \mathrm{d}\boldsymbol{x}(\square + m_\pi^2)\,\bar{D}^\alpha(x) \tag{2.3}$$

we easily find the matrix elements:

$$\langle b|\bar{Q}_0^\alpha|a\rangle = \frac{(2\pi)^3 f_\pi\,\delta(\boldsymbol{p}_b - \boldsymbol{p}_a)\,\langle b|\chi^\alpha|a\rangle}{i(E_b - E_a)}$$
$$\langle b|\bar{Q}_{L,R}^\alpha|a\rangle = \frac{(2\pi)^3 m_\pi f_\pi\,\delta(\boldsymbol{p}_b - \boldsymbol{p}_a)\,\langle b|\chi^\alpha|a\rangle}{i(E_b - E_a)(E_b \pm m_\pi - E_a)} \tag{2.4}$$

where we have introduced the virtual pion source χ^α:

$$(\square + m_\pi^2)\,\bar{D}^\alpha(x) = m_\pi^2 f_\pi\chi^\alpha(x)\,,$$
$$\langle 0|\chi^\alpha|\pi\rangle = \langle\pi|\chi^\alpha|0\rangle = 0\,. \tag{2.5}$$

These simple relations enable us to write down representations for pion amplitudes. As an example let us concentrate on the process $a + V \rightarrow \pi^\alpha + b$ where V can be taken to be an external source, such that $\langle 0|V|\pi\rangle = \langle\pi|V|0\rangle = 0$, and $|a\rangle$, $|b\rangle$ are the initial and final "target" states. The starting point is represented by the introduction of the equal time commutators:

$$\langle b|[\bar{Q}_L^\alpha, V]|a\rangle = \langle b|E_\alpha|a\rangle + (i/m_\pi)\,\langle b|X_\alpha|a\rangle\,, \tag{2.6}$$

where
$$E_\alpha = [\bar{Q}^\alpha, V]\,; \qquad X_\alpha = [\dot{\bar{Q}}^\alpha, V] = O(m_\pi^2)\,.$$

We then insert in Eq. (2.6) a complete set of physical states, make use of the cluster decomposition (see Appendix) and, selecting the pion state, find the sum rule[3]:

$$\langle b|E_\alpha|a\rangle + (i/m_\pi)\,\langle b|X_\alpha|a\rangle$$
$$= \sum_\pi \langle 0|\bar{Q}_L^\alpha|\pi^\beta\rangle\,\langle\pi^\beta, b|V|a\rangle^{\mathrm{conn}}$$
$$+ \sum_{\mathcal{N}} [\langle b|\bar{Q}_L^\alpha|\mathcal{N}\rangle\,\langle\mathcal{N}|V|a\rangle - \mathrm{c.t.}] \tag{2.7}$$
$$+ \sum_{n\ne\mathcal{N},\pi} \left[\frac{(-im_\pi f_\pi)(2\pi)^3\,\delta(\boldsymbol{p}_b - \boldsymbol{p}_n)\,\langle b|\chi^\alpha|n\rangle\,\langle n|V|a\rangle}{(E_b - E_n)(E_b + m_\pi - E_n)} - \mathrm{c.t.}\right],$$

---

[3] From the saturation of axial charge commutators we get, in general, sum rules in which combinations of different pion amplitudes appear. It is easily realized that the "charges" defined in Eq. (2.1), besides having the symmetry breaking already incorporated, enable us to obtain, after use of completeness, a single pion amplitude.

$\sum\limits_{\mathcal{N}}$ representing the contributions of the intermediate states which are degenerate with the external ones (if they are present). In practice the E_α term will consist of a well-known current algebra commutator (finite in the limit $m_\pi \to 0$), plus eventually higher ones, contributing an $O(m_\pi)$ to Eq. (2.7); X_α will be again a higher commutator, affecting the $O(m_\pi)$ part of Eq. (2.7). Of course, all higher commutators are outside the current algebra frame, and further, more specific hypotheses are needed in order to have information about them.

Once we assume the l.h.s. vertex functions to be known, we are then led to the following representation for the *physical pion* amplitude $T_{a+V\to\pi^\alpha+b}$, with the pion at rest[4]:

$$T_{a+V\to\pi^\alpha+b} \equiv i\langle\pi^\alpha, b|V|a\rangle = (1/f_\pi)\langle b|E_\alpha|a\rangle$$
$$+ (i/m_\pi f_\pi)\langle b|X_\alpha|a\rangle - (1/f_\pi)\sum_{\mathcal{N}}[\langle b|\bar{Q}^\alpha_L|\mathcal{N}\rangle\langle\mathcal{N}|V|a\rangle - \text{c.t.}]$$
$$+ m_\pi \int [q_0(q_0-m_\pi)]^{-1}\,dq_0\,\varrho^\alpha(q_0, \boldsymbol{q}=0),\qquad(2.8)$$
$$\varrho^\alpha(q) = i(2\pi)^3\sum_{n\neq\mathcal{N},\pi}\delta(p_b+q-p_n)\langle b|\chi^\alpha|n\rangle\langle n|V|a\rangle - \text{c.t.}$$

Eq. (2.8) is the general formula we shall use in the following, when treating pions at rest; so, let us briefly illustrate its most interesting features, referring to Refs. [1, 2] for more details.

A very important point to be stressed is the analogy between Eq. (2.8) and a dispersion relation, with both energy and "mass" varying, subtracted at the point $q_0 = 0$, the subtraction constant being specified by the equal-time term and the continuum integral contributing on $O(m_\pi)$[5].

As it is easily ascertained, energy and "mass" vary along a certain line, connecting the "soft pion point" ($\boldsymbol{q} = 0$, $m_\pi = 0$) to a "physical point" ($\boldsymbol{q} = 0$, m_π). Our Eq. (2.8) is thus to be regarded as a possible prescription to extrapolate soft-pion theorems, rigorously valid in the limit of vanishing pion four-momentum, where the amplitude is simply expressed by a current algebra commutator plus the nucleon term, into the physical region. The outcome, of course, can be written in the form $T(m_\pi) = T(\text{soft}) + O(m_\pi)$, and the nature of the corrective terms is explicitly displayed. We have, as already anticipated, higher commutators, in-

---

[4] This is due to the fact that a charge, being an integrated operator, can be connected only with pions of vanishing three-momentum.

[5] This is not the only analogy suggested by Eq. (2.8). Remembering Low's treatment of pion physics, still based on equal-time commutators and completeness [3], we can see that Eq. (2.8) is substantially a Low representation for the amplitude $T_{a+V\to\pi^\alpha+b}$, where the identification $\phi_{\pi^\alpha}\leftrightarrow\bar{Q}^\alpha$ has been made, and the subtraction constant specified. Another remarkable resemblance can be established with low-energy nuclear physics, where one has the potential part plus the resonance scattering: it can be shown indeed that the soft-pion amplitude is the analogue of the Born approximation in potential scattering.

volving time derivatives of the axial charge, which appear since the corrections to the symmetry limit $m_\pi = 0$ depend on the way the symmetry is broken, i.e. the pion acquires a finite mass. A further $O(m_\pi)$ is given by the continuum integral, which represents the higher states contributions. Actually, it must be emphasized that such an $O(m_\pi)$ might be, as a matter of fact, an $O(m_\pi(\ln m_\pi)^n)$, due to possible infrared divergences of the integral (see also Ref. [4]). In other words, analytical dependence of Eq. (2.8) on m_π is not granted, and, consequently, simple power-counting in m_π would not be meaningful: the only general statement is that the "corrections" to the soft-pion theorems must vanish as $m_\pi \to 0$. Furthermore, the evaluation of the continuum is in general a difficult task since the vertices $\langle b|\chi^\alpha|n\rangle \equiv F^\alpha((E_b - E_n)^2)$ which appear in ϱ^α are energy-dependent and, moreover, at timelike momentum transfer. In practice however, as a first approximation, this dependence can be neglected: the feeling is that such an approximation affects only the less important part of the amplitude.

The next fundamental point concerns the choice of a reference frame. We are working out a configuration where the pion is at rest, but p_a, p_b are still arbitrary and, in principle, any point of the physical region can be reached. In other words, the sum rule Eq. (2.8) is frame-dependent and we can take advantage of this to enhance (or to depress) certain contributions with respect to others. We will choose nearly rest frame saturation. The reason is that in the configuration $p_a = p_b = 0$ a selection rule holds, due to parity and angular momentum conservation, which strongly reduces the spectrum of intermediate states, allowing only s-waves at the vertices $\langle b|\bar{Q}_L^\alpha|n\rangle$ and $\langle n|\bar{Q}_L^\alpha|a\rangle$.

In general higher waves will be depressed if both $|p_a|$ and $|p_b|$ are small. We obtain thus a very neat separation between the s-wave and the p-wave (and higher), whose presence is eventually related to small (nonvanishing) values of p_a, p_b. While keeping this general limitation in mind, we can choose the reference frame at our convenience, depending on the process under investigation. In the pion-nucleon scattering the choice is immediate, since the configuration $p_a = p_b = 0$ is allowed in this case, and we reach, of course, the physical threshold. In the electroproduction process, owing to the more complicated kinematics, many possibilities are available, among which the Breit frame $p_b = -p_a = p$ will be chosen as the most convenient one, for symmetry reasons.

Our attitude, in conclusion, is to look for a configuration of the external particles, such as to have the simplest "correction" terms and a good control on them; in so doing the "best" extrapolation will be obtained, in the sense that the final result is believed to be as close as possible to the soft pion limit $m_\pi = 0$. From the point of view of the $SU(2) \times SU(2)$ symmetry this evidently corresponds to enhance the

Goldstone mechanism (massless pions, isospin multiplets) with respect to the other possible ones.

The best illustration of the general ideas discussed so far is perhaps their application to specific pion processes. Let us devote then the second part of this Section to pion-nucleon scattering at threshold, reserving Sections III, IV to the photo- and electroproduction processes.

b) Pion-Nucleon Scattering

We have to take in Eq. (2.8) $|a\rangle$, $|b\rangle$ nucleon states with momenta p_1, p_2 and $V \equiv \bar{Q}_0^\beta$.

The pion term is then:

$$\frac{1}{2}(2\pi)^{-3} \int d\pi \cdot E_\pi^{-1} \langle 0|\bar{Q}_L^\alpha|\pi\rangle \langle \pi, p_2|\bar{Q}_0^\beta|p_1\rangle$$
$$= (f_\pi^2/m_\pi)(2\pi)^3 \delta(\boldsymbol{p}_2 - \boldsymbol{p}_1) T_{\pi N}^{\alpha\beta} \tag{2.9}$$

where $T_{\pi N}^{\alpha\beta} = \langle \pi^\alpha, p|\chi^\beta|p\rangle$ is the forward pion nucleon amplitude with the pion at rest. We specialize moreover to the nucleons rest frame $\boldsymbol{p} = 0$, reaching thus the physical threshold configuration, and introduce the s-wave scattering lengths $a^{\alpha\beta}$:

$$T_{\pi N}^{\alpha\beta}(\text{th}) = 8\pi(M + m_\pi) a^{\alpha\beta}. \tag{2.10}$$

The current algebra information (or, if one prefers, the soft-pion result) is contained in the well-known $SU(2) \times SU(2)$ commutator

$$\langle p_2|[\bar{Q}^\alpha, \bar{Q}^\beta]|p_1\rangle = i\varepsilon_{\alpha\beta\gamma}\langle p_2|Q^\gamma|p_1\rangle$$
$$= i\varepsilon_{\alpha\beta\gamma}(\tau^\gamma/2)(2\pi)^3 2E\delta(\boldsymbol{p}_2 - \boldsymbol{p}_1) \tag{2.11}$$

and, in addition, the higher commutators $\langle p|[\dot{\bar{Q}}, \dot{\bar{Q}}]|p\rangle^6$, $\langle p|[\bar{Q}, \ddot{\bar{Q}}]|p\rangle$ and $\langle p|[\ddot{\bar{Q}}, \dot{\bar{Q}}]|p\rangle$ are required.

It is convenient to introduce the decomposition into even and odd part (under $\alpha \leftrightarrow \beta$):

$$T^{\alpha\beta} = \delta_{\alpha\beta} T^{(+)} + \frac{1}{2}[\tau^\alpha, \tau^\beta] T^{(-)}. \tag{2.12}$$

Collecting then Eqs. (2.8)–(2.12) we find very easily

$$(1 + m_\pi/M) a^{(-)} = L(1 + \delta a^{(-)}), \tag{2.13}$$

$$(1 + m_\pi/M) a^{(+)} = L\delta a^{(+)}, \tag{2.14}$$

---

[6] Between states of equal four-momenta, $\langle[\dot{\bar{Q}}, \ddot{\bar{Q}}]\rangle = -\langle[\ddot{\bar{Q}}, \dot{\bar{Q}}]\rangle$.

where $a^{(-)} = \frac{1}{3}(a_1 - a_3)$, $a^{(+)} = \frac{1}{3}(a_1 + 2a_3)$, and $L = m_\pi/8\pi f_\pi^2 \sim 0.09\, m_\pi^{-1}$ is a universal length, independent of the mass of the target. Moreover[7]

$$\delta a^{(-)} = (2m_\pi^2 f_\pi^2/\pi M) \int\limits_{m_\pi}^{\infty} \frac{dq_0\, \varrho^{(-)}(q_0, \mathbf{q} = 0)}{q_0^2(q_0^2 - m_\pi^2 - i\varepsilon)} + m_\pi^{-2} \langle [\dot{\bar{Q}}, \dot{Q}] \rangle, \tag{2.15}$$

$$\delta a^{(+)} = (2m_\pi^2 f_\pi^2/\pi M) \int\limits_{m_\pi}^{\infty} \frac{dq_0\, \varrho^{(+)}(q_0\, \mathbf{q} = 0)}{q_0(q_0^2 - m_\pi^2 - i\varepsilon)} \tag{2.16}$$

$$+ m_\pi^{-1} \langle [\bar{Q}, \dot{Q}] \rangle + m_\pi^{-3} \langle [\dot{\bar{Q}}, \ddot{Q}] \rangle,$$

$$\varrho^{\alpha\beta}(q) = \frac{1}{2} \int d^4x\, e^{iqx} \langle p | [\chi^\alpha(x), \chi^\beta(0)] | p \rangle; \quad \mathbf{p} = 0. \tag{2.17}$$

We have thus a theoretical prediction for the pion-nucleon scattering lengths, in terms of equal-time commutators and a continuum (only s-wave excitation!); it reduces of course to the well-known Weinberg-Tomozawa [5] result in the limit $m_\pi \to 0$[8,9].

Actually pion-nucleon is a very unlucky case, because, while we can be confident in an approximate evaluation of the continua, which can be handled in a relatively simple way, the nature of the higher commutators, on the other hand, is quite complicated. Apart from the σ-term $\langle N | [\dot{\bar{Q}}, \bar{Q}] | N \rangle$, which is very interesting since it gives information about the transformation properties of the symmetry breaking, we need the $[\bar{D}, \bar{D}]$ and $[\bar{D}, \dot{D}]$ commutators, on which we cannot make any general statement at all. We are inclined to use field-algebra like models, since the present treatment of the low energy region is roughly similar to a Lagrangian one, once the identification $\phi_{\pi^\alpha} \leftrightarrow \bar{D}^\alpha$ is made. In so doing we can rely on canonical quantization rules[10], and hope to get rid of both $[\bar{D}, \bar{D}]$ and $[\bar{D}, \dot{D}]$ (actually, the latter is interaction dependent). In this respect the electroproduction process is much simpler, since, as we shall see, the problem is particularly insensitive to higher commutators (involving the breaking).

---

[7] Owing to the selection rule introduced in Part a), the nucleon cannot contribute.

[8] We remind that it may be misleading to ascribe a simple power dependence on m_π to the corrections. Logarithmic dependences, or even a lowering of the power are not excluded.

[9] It is perhaps worth to mention the consistency condition we get from comparing with dispersion relations:

$$\nu^{-1} T^{(-)}(\nu)|_{th} = f_\pi^{-2} + O(m_\pi^2) = g_{\pi N}^2/(M^2 - (m^4/4\, M^2)) + (2/\pi) \int\limits_0^{\infty} \frac{d\nu}{\nu^2 - M^2 m_\pi^2}\, \mathrm{Im}\, T_{(\nu)}^{(-)}.$$

Performing the limit $m_\pi \to 0$, and using the Goldberger-Treiman relation, the Adler-Weisberger sum rule follows:

$$r_A^{-2} = 1 + (2\, M^2/g_{\pi N}) \int \nu^{-1} d\nu\, [\sigma(\pi^- p) - \sigma(\pi^+ p)].$$

[10] This is impossible in other models, e.g. in the quark-model.

If we try anyway a comparison with the experimental results[11a], forgetting the correction effects, we find a very good agreement:

$$(a_1)^{\text{theory}} \approx 0.166 \, m_\pi^{-1} \qquad (a_1)^{\text{exp}} = (0.18 \pm 0.005) \, m_\pi^{-1}$$
$$(a_3)^{\text{theory}} \approx -0.083 \, m_\pi^{-1} \qquad (a_3)^{\text{exp}} = -(0.088 \pm 0.004) \, m_\pi^{-1} . \qquad (2.18)$$

As far as the continuum corrections[11b] are concerned, the complete structure of the spectral functions $\varrho^{(+), (-)}$ is easily established, by using the cluster decomposition, and the result is $\delta a = \delta a_{\text{I}} + \delta a_{\text{II}} + \delta a_{\text{III}}$, where $\delta a_{\text{I, II, III}}$ represent, respectively, the contributions of connected, semidisconnected and Z-graphs. Detailed calculations of these terms are not feasible, but, however, indications about their order of magnitude can be obtained, taking into account the structure of the integrals in Eqs. (2.15), (2.16) and the masses of the lowest isobars [2]. It turns out that, in practice, it should be reasonable to keep only the leading contribution to δa_{I}, i.e. the $\pi - N$ s-wave intermediate state (only $\frac{1}{2}^-$ states are allowed in δa_{I}), and assume the integrals to be dominated by the threshold region. The outcome is, using elastic unitarity

$$\delta a^{(-)} = (2 m_\pi / 3 \pi L) (a_1^2 - a_3^2)$$
$$\delta a^{(+)} = (m_\pi / 3 L) (a_1^2 + 2 a_3^2) . \qquad (2.19)$$

Inserting the experimental numbers, this is seen to affect very slightly the soft-pion predictions quoted in (2.18).

An alternative attitude could be to try to infer, from the knowledge of the experimental data and from an approximate evaluation of the continua, information about higher commutators; in particular, about the σ-term. This problem has recently received much attention, and is really a controversial question. So, although not willing to enter into details, let us end this Section by briefly commenting on it.

The first indication came from Kim and Von Hippel [6] (see also Refs. [7, 8]), who generalized the above approach to meson-baryon scattering at threshold, in the framework of the $SU(3) \times SU(3)$ description[12]. The role played by the continua appears to be really crucial in this case; drastic approximations and uncertain experimental data may cast doubts on their calculations. However, if we believe in their results, we have indications that the $(\bar{3}, 3) + (3, \bar{3})$ model of $SU(3) \times SU(3)$ symmetry breaking [9] fits, and, furthermore, $\sigma \sim 30$ MeV. This is of course the size we expect if $SU(2) \times SU(2)$ has to be a better symmetry (namely, less broken) than $SU(3)$ (remember that $\sigma \sim m_\pi^2$).

[11a] References and values based on more recent data are given in the compilation of Ebel et al. [Nucl. Phys. B**22**, 317 (1971) (Ed.)].

[11b] Strictly speaking, the term "corrections" should apply to $a^{(-)}$ only, whereas $a^{(+)}$ is given by a sum of terms which are formally of the same order, as is evident from Eqs. (2.13)–(2.16).

[12] We refer to Ref. [9] for what concerns the general scheme of $SU(3) \times SU(3)$.

On the other hand, an alternative approach to the σ-term, based on fixed t dispersion relations for the pion-nucleon isospin-symmetric amplitude $T_{\pi N}^{(+)}$, seems to indicate a quite larger value, namely $\sigma \sim 100$ MeV [10] (which, by the way, is the order of magnitude of the $SU(3)$ breaking). The assumptions and the hypotheses made in this kind of calculations are of different nature, but the reliability of the results is certainly not improved, since, among other things, reasonable uncertainties of the experimental input lead to strong variations of the final outcome[13]. Anyway, a large value of the σ-term would be perhaps welcome by theorists (actually, it has been, and papers on that are becoming more and more numerous), since it could be a fruitful starting point for further speculations concerning chiral symmetry. Among the many promising ideas let us mention, for instance, the possible connection with a (spontaneously) broken dilatation symmetry[12]. This is why the problem of the σ-term, which, in our opinion, is still an open question, deserves further investigations: a better understanding of chiral symmetry and its breaking will perhaps result.

As a final remark, let us emphasize that the approach we have presented here to perform the extrapolation from soft to physical pions, is, in some way, the most economical and nearest to the original current algebra framework. In particular the minimal number of commutators, beyond the ones of the $SU(3) \times SU(3)$ algebra, has been introduced: a prize to pay for this is the very nature of the continua. It is clear that an improved knowledge of higher commutators and a more detailed information about them would enable one to have more and more sum rules, from which to eliminate the unpleasant contributions. We have in mind recent approaches to the extrapolation problem, starting from the knowledge of light-cone commutators: such an information (that roughly amounts to an infinite number of equal time commutators) seems to allow a different representation of the continuum corrections, but their estimate is anyway not easy [13].

III. Electroproduction at Threshold

As a further example, we now consider the electro- (and photo-) production of a single pion. This process has represented a test of very different approaches, and to try to review the large number of related papers would ask for a separate effort [14]. From our point of view it will be enough to recall the relevant "soft pion theorems". In photoproduction

---

[13] The dispersive evaluation of Ref. [11], for instance, yields the result $\sigma \sim 40$ MeV.

Note added in proof. The result of Ref. [11] is essentially confirmed by a calculation, in which the new CERN phase shifts (Oct. 71) were used (H. P. Jakob, CERN preprint TH 1446) (Ed.).

the standard application of a method[14] due to Low [15] leads to the so-called Kroll-Ruderman [16] theorem for the s-wave multipoles, stating that:

$$\lim_{m_\pi \to 0} E^{(-)}_{0+\,(\text{th.})} = (1/2M)\, g_{\pi N}(m_\pi = 0)\,,$$

$$\lim_{m_\pi \to 0} E^{(0),(+)}_{0+\,(\text{th.})} = 0\,, \tag{3.1}$$

where the usual isotopic decomposition of the photoproduction amplitude

$$T^\alpha_\mu = \delta_{\alpha 3}\, T^{(+)}_\mu + \tfrac{1}{2}\,[\tau^\alpha, \tau^3]_-\, T^{(-)}_\mu + \tau_\alpha\, T^{(0)}_\mu\,. \tag{3.2}$$

has been used.

When the photon is off mass shell, we have the following formulae, derived by Nambu and coworkers [17]:

$$E^{(-)}_{0+}(\text{S.P.L.}) = \sqrt{\frac{4M^2 - \kappa^2}{4M^2}}\,\frac{g_{\pi N}(0)}{2M}\left[\frac{G_A(\kappa^2)}{G_A(0)} + \frac{\kappa^2}{4M^2 - 2\kappa^2}\,G^v_M(\kappa^2)\right], \tag{3.3}$$

$$E^{(0),(+)}_{0+}(\text{S.P.L.}) = \sqrt{\frac{4M^2 - \kappa^2}{4M^2}}\,\frac{g_{\pi N}(0)}{2M}\left[\frac{-\kappa^2\, G^{s,v}_M(\kappa^2)}{4M^2 - 2\kappa^2}\right], \tag{3.4}$$

$$\mathscr{L}^{(-)}_{0+}/\kappa_0(\text{S.P.L.}) = \sqrt{\frac{4M^2 - \kappa^2}{4M^2}}\,g_{\pi N}(0)\left[\frac{G^v_E(\kappa^2)}{4M^2 - 2\kappa^2}\right], \tag{3.5}$$

$$\mathscr{L}^{(0),(+)}_0/\kappa_0(\text{S.P.L.}) = \sqrt{\frac{4M^2 - \kappa^2}{4M^2}}\,g_{\pi N}(0)\left[\frac{-G^{s,v}_E(\kappa^2)}{4M^2 - 2\kappa^2}\right] \tag{3.6}$$

and for a quick derivation of them, which also summarizes the main features of the fundamental work of Nambu, we refer to a previous paper [18]. In Eqs. (3.3)–(3.6) and (3.1) E_{0+}, $\mathscr{L}_{0+}$ are the s-wave multipoles, defined as in Ref. [18], and a definition of the different form factors entering the various expressions will be found in going through this paper.

A non trivial point, which has been first emphasized by Nambu, is the frame-dependence which the soft pion limits may in general exhibit. From a point of view of dispersion relations, this happens as a rule for all those quantities which receive contribution from the Born approximation. Actually, a prescription to reach the situation where the pion has both vanishing mass and momentum has to be specified. Usually, one means by "soft pion limit" the limit attained by setting first the pion three-momentum, and then the pion mass, equal to zero. In this way, a frame is automatically chosen; in particular, Eqs. (3.3)–(3.6) are derived by this limiting procedure in the final pion-nucleon center-of-mass

[14] A rigori, this method would lead in general to "soft photon" theorems. However, if the photon is real, it reduces to the form Eq. (3.1).

frame, where the multipoles are defined. (This should be kept in mind if one wanted to compare our final formulae with Eqs. (3.3)–(3.6)).

To begin with, we consider the simple configuration where the final pion is at rest. The relevant commutator, to be inserted between two one-nucleon states of three-momentum p_2, p_1 respectively, is then:

$$[\bar{Q}_L^\alpha, V_\mu] = E_\mu^\alpha . \tag{3.7}$$

Here V_μ is the electromagnetic current, $V_\mu = V_\mu^{(3)} + V_\mu^{(0)}$ and E_μ^α is the sum of a standard charge-current commutator

$$[\bar{Q}^\alpha, V_\mu] = i\varepsilon_{\alpha 3\gamma} A_\mu^\gamma \tag{3.8}$$

and of a term explicitly depending on the breaking.

$$[\dot{\bar{Q}}^\alpha, V] = \int d\boldsymbol{x}[\bar{D}^\alpha(\boldsymbol{x}, t), V] \tag{3.9}$$

(for the time component the relation $[\dot{\bar{Q}}^\alpha, V_0] = i\varepsilon_{\alpha 3\gamma}\bar{D}^\gamma$ follows immediately from Eq. (3.8) by the electromagnetic current conservation).

We do not make at the moment any assumption about the commutator $[\bar{D}^\alpha, V]$. Its role in the determination of the s-wave multipoles at threshold is practically irrelevant, whilst it has some importance in the case of certain higher multipoles. Therefore, in the first part of this Section, let us simply ignore it.

To proceed on, it is necessary to fix the momenta of the external nucleons, i.e. to choose a definite frame. For a series of reasons partially explained in the previous Section, it seems to be very convenient to choose the Breit frame of the two external nucleons, $p_1 + p_2 = 0$.

Following the rules of the game illustrated in the previous Section, we then select the contribution coming from the one pion state and are led in a straightforward way to the expression

$$T_\mu^\alpha(v, t, \kappa^2)_{\text{B.t.}} = \frac{1}{f_\pi}\left[\langle p_2|E_\mu^\alpha|p_1\rangle - \sum_{n \neq \pi}\langle p_2|\bar{Q}_L^\alpha|n\rangle\langle n|V_\mu|p_1\rangle + \text{c.t.}\right] \tag{3.10}$$

where T_μ^α is the physical single pion electroproduction amplitude

$$T_\mu^\alpha \equiv i\langle N(p_2), \pi^\alpha(q=0)|V_\mu|N(p_1)\rangle \tag{3.11}$$

evaluated in the kinematical configuration where the pion is at rest in the Breit frame, which means, in a covariant notation, at the point:

$$\begin{aligned} v &= m_\pi E \equiv v_{\text{B.t.}} \\ t &= \kappa^2 - m_\pi^2 \equiv t_{\text{B.t.}} \end{aligned} \tag{3.12}$$

where $(p_1 + \kappa = p_2 + q)$:

$$\begin{aligned} v &= P\cdot\kappa = P\cdot q ; \quad t = (p_2 - p_1)^2 \\ P &= \tfrac{1}{2}(p_1 + p_2) ; \quad P^2 = M^2 - t/4 ; \quad E \equiv \sqrt{P^2} . \end{aligned} \tag{3.13}$$

Although this is not strictly necessary, it is convenient at this point to adopt a fully non-covariant notation and effectively evaluate Eq. (3.10) in the frame $P = 0$. To this purpose we need the explicit expressions of the nucleon-axial current and nucleon electromagnetic vertices, which are respectively

$$\langle p | A^\alpha | - p \rangle = (\tau^\alpha/2) \{ 2 E G_A(t) [n(\sigma \cdot n) - \sigma] + n\sigma \cdot n D(t) \},$$
$$\langle p | A_0^\alpha | - p \rangle = 0 ; \quad \langle p | \bar{D}^\alpha | - p \rangle = (\tau^\alpha/2 i) \, \sigma \cdot \kappa D(t),$$

(3.14)

where

$$D(t) = - 2 M G_A(t) + t G_P(t) ; \quad n \equiv \kappa/|\kappa| \tag{3.15}$$

and

$$\langle p | V_0^{(3),(0)} | - p \rangle = \tfrac{1}{2} \langle \tau^3, 1 \rangle 2 M G_E^{v,s}(t) \tag{3.16}$$

$$\langle p | V^{(3),(0)} | - p \rangle = \tfrac{1}{2} \langle \tau^3, 1 \rangle i\sigma \times \kappa G_M^{v,s}(t), \tag{3.17}$$

where $G_{E,M}$ are the Sachs form factors:

$$G_M = F_1 + F_2 ; \quad G_E = F_1 + (t/4M^2) F_2 . \tag{3.18}$$

Concerning the amplitude T_μ^α, we choose its space components as follows

$$T = A \sigma + B[\kappa(\sigma \cdot \kappa) - \kappa^2 \sigma] \tag{3.19}$$

and *define* its time component through the gauge-invariance condition:

$$\kappa \cdot T = \kappa_0 T_0 (\kappa_0 = m_\pi). \tag{3.20}$$

By comparing the time and space components of Eq. (3.10) it is now easy to deduce the equations, where a "real part" is always understood:

$$A^{(-)} = D(t)/2 f_\pi + \delta A^{(-)}, \tag{3.21}$$

$$A^{(+),(0)} = \frac{m_\pi M}{2 E f_\pi} G_A(0) G_E^{v,s}(t) + \delta A^{(+),(0)}, \tag{3.22}$$

$$T_T^{(-)} \equiv \frac{1}{2E} (A + tB)^{(-)} = - \frac{G_A(t)}{2 f_\pi}$$
$$- \frac{t G_A(0) G_M^v(t)}{8 E^2 f_\pi} + \frac{1}{2E} [\delta A + t\delta B]^{(-)}, \tag{3.23}$$

$$T_T^{(+),(0)} = m_\pi M G_A(0) G_M^{v,s}(t)/4 E^2 f_\pi$$
$$+ t m_\pi \tilde{B}^{(+),(0)}/2 E + \frac{1}{2E} [\delta A + t\delta B]^{(+),(0)} \tag{3.24}$$

where

$$\delta(A^{(-)}, B^{(-)}) = 2m_\pi^2 \int_0^\infty \frac{dq_0 \varrho_{A,B}^{(-)}(q_0, t)}{q_0(q_0^2 - m_\pi^2 - i\varepsilon)}, \tag{3.25}$$

$$\delta A^{(+),(0)} = 2m_\pi^3 \int_0^\infty \frac{dq_0 \varrho_A^{(+),(0)}(q_0, t)}{q_0^2(q_0^2 - m_\pi^2 - i\varepsilon)}, \tag{3.26}$$

$$\delta B^{(+),(0)} = 2m_\pi \int_0^\infty \frac{dq_0 \varrho_B^{(+),(0)}(q_0, t)}{(q_0^2 - m_\pi^2 - i\varepsilon)}. \tag{3.27}$$

$\varrho_{A,B}$ are the components (corresponding to the decomposition Eq. (3.19)) of

$$\varrho^\alpha = i(2\pi)^3 \sum_{n \neq \pi, \mathcal{N}} \delta(p + q - p_n) \langle p | \chi^\alpha | n \rangle \langle n | V | - p \rangle \tag{3.28}$$

and the quantity $m_\pi \tilde{B}^{(+),(0)}$ is the contribution to $B^{(+),(0)}$ from the commutator $i/m_\pi \langle p | [\dot{Q}^\alpha, V] | - p \rangle$ (it does not contribute to the real part of $B^{(-)}$ owing to the crossing properties of ϱ_μ^α).

The reason for the choice of the particular combinations (3.21)–(3.24) is that, following the notation commonly adopted in the literature [19], one can easily verify that (we are now in the c.m.s., where W and θ_π are the total energy and the pion angle respectively):

$$(|\kappa|/|q|)(d\sigma_T/d\cos\theta_\pi)_{B.t.} = (\alpha E^2/2W^2)/|T_{T\,B.t.}|^2, \tag{3.29}$$

$$(|\kappa|/|q|)(d\sigma_L/d\cos\theta_\pi)_{B.t.} = -(\kappa^2\alpha/8W^2 m_\pi^2)|A_{B.t.}|^2. \tag{3.30}$$

According to the remarks of the previous section, it is clear that the still not evaluated terms in Eqs. (3.21)–(3.24) will be vanishing together with the pion mass. If we neglect for the moment the fact that the actual corrections can lead to a pion mass dependence of the corrections different from the explicit powers multiplying the continuum integrals, then, in the specific case of photoproduction, we can summarize our results in the simple form:

$$T_T^{(-)}(\kappa^2 = 0; B.t.) = -\frac{G_A(0)}{2f_\pi} + O(m_\pi^2). \tag{3.31}$$

$$T_T^{(+),(0)}(\kappa^2 = 0; B.t.) = \frac{m_\pi G_A(0)}{4Mf_\pi} + O(m_\pi^3). \tag{3.32}$$

Introducing the physical amplitudes

$$T_\pm = \sqrt{2}[T^{(-)} \pm T^{(0)}]; \quad \begin{matrix} \gamma p \to \pi^+ n \\ \gamma n \to \pi^- p \end{matrix}$$

$$T_{\substack{01 \\ 02}} = T^{(+)} \pm T^{(0)}; \quad \begin{matrix} \gamma p \to \pi^0 p \\ \gamma n \to \pi^0 n \end{matrix} \tag{3.33}$$

we find:

$$(|\kappa|/|q|)\, d\sigma_+/d\Omega_{\text{B.t.}} = 12.7\,\mu\text{b/sr} + O(m_\pi^2); \quad \text{Exp.: } 15.6 \pm 0.5^{(20)}$$

$$R \equiv d\sigma_-/d\sigma_+ = 1.35 + O(m_\pi^2); \qquad \text{Exp.: } 1.265 \pm 0.075^{(21)}$$

$$\sim 1.35^{(22)} \qquad\qquad (3.34)$$

$$(|\kappa|/|q|)d\sigma_{01}/d\Omega_{\text{B.t.}} = 0.04\,\mu\text{b/sr} + O(m_\pi^2); \quad \text{Exp.: } 0.07 \pm 0.02^{(23)}$$

$$(|\kappa|/|q|)d\sigma_{02}/d\Omega_{\text{B.t.}} = O(m_\pi^2)$$

whilst the corresponding predictions of the Kroll-Ruderman theorem are respectively[15] : 19,5; 1; 0; 0. Since it is immediate to verify that the Kroll-Ruderman theorem is actually reproduced by Eqs. (3.31), (3.32), we can look at them as a very simple and meaningful generalization of that theorem. The terms respectively classified as $O(m_\pi^2)$ and $O(m_\pi^3)$ are therefore, as we expected, "corrections" (unambiguously defined) to the soft pion theorem Eq. (3.1) (although the final dependence on m_π can turn out to be different, in particular $\ln m_\pi$ can arise).

In a perfectly similar way, the analogous terms in the general equations (3.21)–(3.24) evaluated at $\kappa^2 \neq 0$ will represent "corrections" to some results, which would be exact in a world of massless pions. The corresponding "soft pion formulae" are different from Nambu's Eq. (3.3) (this is not surprising if we remember that the pion has now been put at rest in the Breit frame, and not in the c.m. frame).

A pleasant feature we want to mention here is that, if one imposes that the same results have to be derived by starting from a dispersion theory approach and by going to the limit ($q = 0$, $m_\pi = 0$) in the Breit frame, then one obtains "consistency conditions" which are just the Fubini, Furlan and Rossetti [24] and Furlan, Jengo and Remiddi [25] sum rules[16]. This is just the analogue of the result shown in the previous Section, the "consistency condition" being in that case the Adler-Weisberger sum rule.

---

[15] We have used the value $G_A(0)/2 f_\pi \simeq 0{,}92/m_\pi$.

[16] The sum rules are

$$G_A(t)/G_A(0) = F_1^v(t) - (2\,Mt/g_{\pi N}\pi) \int_{\text{cont}} d\nu \cdot \nu^{-1} \operatorname{Im} M_6^{(-)}(\nu, t, q^2 = 0, \kappa^2 = t), \qquad \text{(I)}$$

$$F_2^{v,s}(t)/2\,M = (2\,M/g_{\pi N}\pi) \int_{\text{cont}} d\nu \cdot \nu^{-1} \cdot \operatorname{Im} M_1^{(+,0)}(\nu, t, q^2 = 0, \kappa^2 = t), \qquad \text{(II)}$$

where $M_{1,6}$ are the invariant coefficients of $\frac{1}{2}\gamma_5[\gamma\cdot\varepsilon, \gamma\cdot k]$, $\gamma_5(\kappa\cdot\varepsilon\gamma\cdot\kappa - \kappa^2\gamma\cdot\varepsilon)$ in the electroproduction amplitude. Usually the sum rules are exploited to get an estimate for $F_2^{v,s}(0)$ and $G_A'(0)$, but one can also read them as soft pion theorems, i.e.

$$M_1^{(+,0)}(\nu, t, q^2 = 0, \kappa^2 = t)|_{\text{cont}} = (g_{\pi N}/2\,M)\, F_2^{v,s}(t), \qquad \text{(III)}$$

$$M_6^{(-)}(\nu, t, q^2 = 0, \kappa^2 = t)|_{\text{cont}} = (g_{\pi N}/2\,Mt)\,[F_1^V(t) - G_A(t)/G_A(0)]. \qquad \text{(IV)}$$

We shall comment on them in the following.

To give a more concrete meaning to Eqs. (3.21)–(3.24) an evaluation of the "corrections" is required. Clearly, if such an evaluation could be exactly performed in terms of known parameters, then from the measurement of electroproduction at the Breit threshold one would be able to learn something on the nucleon axial vector form factors, which contribute to the ($-$) isotopic amplitudes, that is to the charged pion cross sections.

Since the terms to be evaluated are related to off mass shell quantities, their treatment requires unavoidably some "smoothness" assumptions. To minimize the role of the extra hypotheses needed, a first prescription is to choose a "small" t configuration (say, $|t| \lesssim 20\, m_\pi^2$). This is suggested by the fact that, at $t = 0$, the selection rule illustrated in the previous Section would forbid p-waves and higher waves to contribute in the direct channel, so that their contribution at small t should be particularly depressed. Correspondingly, we shall obtain predictions for electroproduction in the range $0 \lesssim |\kappa^2| \lesssim 20 m_\pi^2$. In this region, an evaluation of the "corrections" has been proposed [26], whose main features can be summarized as follows:

a) The contribution of the extra terms is a significant fraction (about 20–30%) of the "leading" (i.e. equal time and one-nucleon) contribution (which would be sufficient to reproduce the "soft pion limit") in the case of the ($-$), (0) amplitudes. However, it appears still licit in this case to treat the extra terms as "small" corrections, to be evaluated by making recourse to a number of unavoidable approximations.

b) The contribution of the extra terms is dramatically larger than the "leading" one for the ($+$) amplitudes. Therefore, all the predictions for neutral pions depend in a rather crucial way on the approximations used to evaluate the unphysical contribution. Owing to this reason, we shall consider the latter predictions as purely indicative, and shall limit ourselves from now on to treat the charged pion processes. Before doing this in more detail, it is fruitful, however, to realize that these formulae, although the simplest, are probably not the most popular ones. First of all they hold at the Breit-threshold (corresponding to pions emitted backward in the c.m.s.) where a small amount of p – and higher waves is present.

A prediction concerning the actual physical threshold would be probably more favourably accepted. Secondly, a generalization from a single-point prediction to a prediction covering a given energy-range would allow a more complete check of the whole approach.

If we do not want to loose the pleasant features of working in the Breit frame, a simple solution is to have predictions for pions which are not at rest in the Breit frame, but have just that amount of three-momentum which corresponds, by a Lorentz transformation, to being at rest in the c.m.s.

This aim can be attained by introducing the operator

$$\mathscr{A}_L^\alpha(q) = A_0^\alpha(q) + i\omega(q)m_\pi^{-2}\,\bar{D}^\alpha(q)\,, \quad \omega = \sqrt{q^2 + m_\pi^2}$$
$$\mathscr{A}_L^\alpha(q) = \int d\mathbf{x}\,\exp(-i\mathbf{q}\cdot\mathbf{x})\,\mathscr{A}_L^\alpha(x)\xrightarrow[q\to 0]{}\bar{Q}_L^\alpha \tag{3.35}$$

with the property

$$\langle 0\,|\,\mathscr{A}_L^\alpha(q)\,|\,\pi^\beta\rangle = 2i\omega f_\pi(2\pi)^3\,\delta_{\alpha\beta}\,\delta(\mathbf{q}-\boldsymbol{\pi})$$
$$\langle\pi\,|\,\mathscr{A}_L^\alpha|\,0\rangle = 0 \tag{3.36}$$

and starting from the commutator:

$$[\mathscr{A}_L^\alpha(q),\,V_\mu(-\mathbf{k})] = E^\alpha + i\omega m_\pi^{-2}\,X^\alpha \tag{3.37}$$

where

$$E^\alpha = [A_0^\alpha,\,V_\mu] \tag{3.38}$$

and

$$X^\alpha = [\bar{D}^\alpha,\,V_\mu]\,. \tag{3.39}$$

As we shall see, we will be able to deduce from these commutators an information for s, p and d-waves which covers a kinematical region ranging from the physical threshold up to the beginning of the first resonance, thus avoiding also the second previously mentioned restriction.

IV. Low Energy Electroproduction and Concluding Remarks

A point to be stressed before we proceed is that if one wants to start from a commutator like Eq. (3.37), involving a charge density rather than a charge, then of course many features of the approach outlined in the previous Sections will change. First of all the chiral symmetry pattern will be partially lost. By this we mean that one will have now two kinds of predictions. The first ones would give back in a particular limit all those results which can be derived by starting from a charge-current commutator as Eq. (3.8); therefore they will still contain a term which reproduces a "soft pion limit" and some "corrections". The second ones are new predictions it is not possible to obtain starting from a charge, and thus they do not correspond to any "soft pion result". Roughly speaking, they involve those invariant amplitudes whose coefficient vanishes in the limit $q_\pi \to 0$.

The amount of new information one obtains in this way must correspond to some new imput, and therefore to a loss of generality of the approach. In our particular case the use of the current-current commutator implies that possible Schwinger terms appear in $[A_0^\alpha, V]$, which are model dependent. In a simple field algebra model which is suggested by the analogy of our approach to a Lagrangian description of low energy pion physics, the mixed Schwinger term in $[A_0^\alpha, V]$ is zero. Other models more generally [27] predict the existence of a gradient term:

$$[A_0^\alpha(x, 0), V_\mu(y, 0)] = i\varepsilon_{\alpha 3 \gamma} A_\mu^\gamma(x)\, \delta(x - y) - i S_{AV}^{\alpha 3}(y)\, g_\mu^\kappa\, \partial_\kappa\, \delta(x - y) \qquad (4.1)$$

provided that:

$$[\bar{D}^\alpha(x, 0), V_0(y, 0)] = i\varepsilon_{\alpha 3 \gamma} \bar{D}^\gamma(x)\, \delta(x - y)\,. \qquad (4.2)$$

In general one can easily verify that, if we saturate Eq. (4.1) in the Breit frame $P = 0$, then this term will contribute to only one appropriately chosen invariant amplitude. But gauge invariance will allow us to replace this amplitude with a combination of the remaining ones, so that within this approach the nature of the Schwinger term in Eq. (4.1) is irrelevant.

It remains to discuss the commutator Eq. (3.39). A pleasant feature of the saturation of densities is that it allows us to see to which multipoles this commutator actually contributes. If we limit ourselves to charged pion processes, i.e., to the amplitudes $(-)$, (0), then it can be shown [18] that this commutator is practically irrelevant to the determination of the s-wave multipoles[17], whilst it will contribute in general to higher waves. Therefore, if we want to push our predictions beyond threshold, we must make some definite hypothesis concerning it. A simple solution, suggested by field theoretical models, is that this commutator is vanishing:

$$[\bar{D}^\alpha, V] = 0\,. \qquad (4.3)$$

We shall accept Eq. (4.3). From the point of view of dispersion theory, it can be seen [18] that this assumption has at least the same validity as the statement that the Born approximation gives a "reasonable" evaluation of the $(-)$, (0) invariant amplitudes at threshold.

The technical details are now not very important and we shall limit ourselves to a sketch of the procedure. The starting point is represented by the Breit frame saturation of the commutator Eq. (3.37). The procedure should be by now familiar, and the result can be written as [28]

$$\begin{aligned} f_\pi T_\mu^\alpha(p, q) = {}& \langle [A_0^\alpha, V_\mu] \rangle \\ & + (i\omega/m_\pi^2) \langle [\bar{D}^\alpha, V_\mu] \rangle + (\text{one nucleon}) \qquad (4.4) \\ & + \omega R_\mu^\alpha + q \cdot S_\mu^\alpha \end{aligned}$$

where $R_\mu^\alpha, S_\mu^\alpha$ are directly connected to completeness sums. The factors ω, q clearly indicate that we are extrapolating a result, originally obtained

---

[17] More precisely, it contributes to a kinematically depressed part of $E_{0+}^{(0)}$, which, in turn, is a small quantity compared with $E_{0+}^{(-)}$.

in the limit q, $m_\pi \rightarrow 0$, to a whole range of the physical region. Eq. (4.4) allows us to derive, after appropriate projections, the complete set of six invariant amplitudes at a certain physical point, specified by q. In particular, we shall choose q so as to obtain that kinematical configuration which corresponds to the physical threshold. The next step is then to transform our language and to introduce the experimentally meaningful multipoles. In the low energy region we want to consider, namely from threshold up to $|q_\pi^{C.M.}| \lesssim m_\pi$ we are allowed to use, for these quantities, power series in $|q_\pi^{C.M.}|$ truncated at the lowest power. Moreover, even if, in principle, Eq. (4.4) could be exploited in a general configuration, we prefer to fix $\theta = 90°$ (in the c.m.s.), for simplicity reasons. We are able thus to describe electroproduction at $\theta = 90°$, in an energy region from threshold up to the starting of the first resonance, and for not too large momentum transfer between the electrons (say, $|\kappa^2| \lesssim 20\, m_\pi^2$).

To be more definite, a number of final formulae is unavoidable. We first write down the expression of the differential cross section, including terms of order $|q_\pi^{C.M.}|^2$:

$$\frac{d^3\sigma}{d\Omega_l'^L \, dE_l'^L \, d\Omega_\pi} = \frac{\alpha^2}{(2\pi)^3} \frac{E_l'^L}{E_l^L} \frac{|\kappa|}{(-\kappa^2)} \frac{M}{W} \frac{1}{1-\varepsilon} \frac{|q_\pi|}{|\kappa|}$$

$$\{[A + B\cos\theta + C\cos^2\theta] + (\varepsilon(-\kappa^2)/\kappa_0^2)[D + E\cos\theta + F\cos^2\delta]$$

$$+ \varepsilon\cos 2\phi \sin^2\theta\, G + \sqrt{2\varepsilon(1+\varepsilon)} (-\kappa^2)/\kappa_0^2 \cos\phi \sin\theta[H + I\cos\theta]\}$$
(4.5)

where E_l^L, $E_l'^L$ are the lepton energies in the laboratory system, $\Omega_\pi \equiv (\theta, \phi)$ is the $\pi - N$ solid angle and ε is the "polarization" defined as:

$$\varepsilon = [1 - (2|\kappa^L|^2/\kappa^2)\, \mathrm{tg}^2\, \theta_l'/2]^{-1}$$

A, B, C are given by the following equations:

$$A = F_1^2 + \tfrac{9}{2}z^2 + y^2 + 3yz + 3\chi\bar{E}_{0+}$$

$$B = 2x\bar{E}_{0+}$$
(4.6)

$$C = -A + F_1^2 + x^2 + 15\bar{E}_{0+}\psi$$

where

$$\bar{E}_{0+} \equiv E_{0+}(|q_\pi| = 0); \quad F_1 \equiv |E_{0+} + 3M_{2-} + E_{2-} - 3M_{2+} - \tfrac{3}{2}E_{2+}|.$$

$$z = \mathrm{Re}(E_{1+} - M_{1+}). \qquad \chi = \mathrm{Re}(M_{2+} - E_{2+} - M_{2-} - E_{2-}).$$

$$y = \mathrm{Re}(2M_{1+} + M_{1-}). \qquad \psi = \mathrm{Re}(2M_{2+} + E_{2+}).$$
(4.7)

$$x = \mathrm{Re}(3E_{1+} + M_{1+} - M_{1-})$$

and for the remaining coefficients we have:

$$D = F_5^2 + v^2; \quad E = 2u\mathscr{L}_{0+}; \quad F = u^2 - v^2 + 45\mathscr{L}_{0+}\mathscr{L}_{2+};$$
$$G = \tfrac{9}{2}z^2 + 3yz + 3\chi \bar{E}_{0+}; \quad H = \mathscr{L}_{0+}(y+3z) + \bar{E}_{0+}v;$$
$$I = u(y+3z) + xv + \bar{E}_{0+}\Xi + \mathscr{L}_{0+}\zeta;$$
$$F_5 = |\mathscr{L}_{0+} + 2\mathscr{L}_{2-} - \tfrac{9}{2}\mathscr{L}_{2+}|; \quad \mathscr{L}_{0+} \equiv \mathscr{L}_{0+}(|\mathbf{q}_\pi| = 0); \tag{4.8}$$
$$u = 4\mathscr{L}_{1+} + \mathscr{L}_{1-}; \quad v = \mathscr{L}_{1-} - 2\mathscr{L}_{1+};$$
$$\Xi = 6\mathscr{L}_{2-} - 9\mathscr{L}_{2+}; \quad \zeta = 12E_{2+} - 3E_{2-} - 3M_{2-} - 3M_{2+}.$$

The quantities which are needed to describe the process in our configuration $\theta = 90°$ can be derived by a lengthy but straightforward computation. Since their explicit expression is rather complicated, we limit ourselves here to give the s-wave multipoles at threshold, and refer to Ref. [18] for a complete list of the results.

$$E_{0+}^{(-)}(\text{th}) = -\sqrt{P^2/M^2}\,(G_A(t)/2f_\pi)|_{\text{th}} - \sqrt{P^2/M^2}\,[4P^2 f_\pi \sqrt{1 - \kappa \cdot q'/P^2}]^{-1}$$
$$\cdot [(\kappa_N^2/2)\,G_A(q_N^2)\,G_M^v(\kappa_N^2) - (M/P^2)\,v_N^2\,G_P(q_N^2)\,G_M^v(\kappa_N^2)]|_{\text{th}} + \delta E_{0+}^{(-)}, \tag{4.9}$$

$$E_{0+}^{(0)}(\text{th}) = \sqrt{P^2/M^2}\,\frac{m_\pi G_A(q_N^2) F_1^s(\kappa_N^2)}{2M\sqrt{1 - \kappa \cdot q'/P^2}} + \sqrt{P^2/M^2}\,[4P^2 f_\pi \sqrt{1 - \kappa \cdot q'/P^2}]^{-1}$$
$$\cdot [G_A(q_N^2) F_2^s(\kappa_N^2)(\kappa^2 P^2/2M^2) + m_\pi v_N(F_1^s(\kappa_N^2) + F_2^s(\kappa_N^2)t/4M^2)] + \delta E_{0+}^{(0)}, \tag{4.10}$$

$$\mathscr{L}_{0+}^{(-)}/\kappa_0|_{\text{th}} = \sqrt{P^2/M^2}\,\frac{D(t)}{2m_\pi f_\pi(2M+m_\pi)} + \sqrt{P^2/M^2}\,[4P^2 f_\pi \sqrt{1-\kappa \cdot q'/P^2}]^{-1}$$
$$\cdot \frac{M\,D(q_N^2)\,G_E^v(\kappa_N^2)}{2M+m_\pi} + \delta\mathscr{L}_{0+}^{(-)}/\kappa_0, \tag{4.11}$$

$$\mathscr{L}_{0+}^{(0)}/\kappa_0|_{\text{th}} = \sqrt{M^2/P^2}\,\frac{G_A(q_N^2)\,G_E^s(\kappa_N^2)}{2f_\pi(2M+m_\pi)} + \sqrt{P^2/M^2}\,[4P^2 f_\pi \sqrt{1-\kappa \cdot q'/P^2}]^{-1}$$
$$\cdot \frac{M}{2M+m_\pi}\left[\left(-\frac{v_N^2}{P^2} + \frac{m_\pi v_N}{M}\right)G_P(q_N^2)\,G_M^s(\kappa_N^2)\right. \tag{4.12}$$
$$\left. - \frac{m_\pi v_N}{M}\,G_P(q_N^2)\,F_2^s(\kappa_N^2)\right] + \delta\mathscr{L}_{0+}^{(0)}/\kappa_0$$

$$\cdot \left(t_{\text{th}} = \frac{M(\kappa^2 - m_\pi^2)}{M+m_\pi}; \kappa_N^2 \simeq t; q_N^2 = O(m_\pi^2); v_N^2 = O(m_\pi^2); \kappa \cdot q' = O(m_\pi)\right).$$

The "corrections" are still vanishing with the pion mass, and one can easily check that Eqs. (4.9)–(4.12) reproduce the "soft pion limits" Eqs. (3.3)–(3.6), of which they are therefore a meaningful and unambiguous extension. The higher waves represent, of course, a new result, in the sense that, in this case, there is no low-energy theorem to extrapolate. We have now to evaluate the "corrections". This evaluation has been performed in previous references [28], and we do not want to quote their complicated expressions here. From a general point of view we can say that there are some "corrections" which can be very reasonably evaluated without making recourse to particular assumptions, i.e. the s-wave continuum and the N^* contribution in the s-channel (the latter being always extremely small). In addition we have to evaluate more complicated terms. Our empirical assumption is that the main contribution comes from the vector meson exchange in the κ^2-channel and that other graphs, e.g. the 3π contributions, can be ignored.

In this way we are led to a representation of the various multipoles containing a "leading" current algebra and one nucleon term and "corrections" whose typical order of magnitude is a 10–20% of the overall quantity. The next important point is then the comparison with experimental data. This may be done only if we choose a definite expression of the axial vector form factors (we suppose that a dipole fit with $M_V = 6m_\pi$ well accounts for the electromagnetic ones). In particular, if we choose the following parametrization (we work in the small t region)

$$G_A(t) = (1 - t/M_A^2)^{-2} G_A(0), \tag{4.13}$$

$$D(t) \simeq 2m_\pi^2 f_\pi g_{\pi N}/(m_\pi^2 - t) \tag{4.14}$$

then, fixing $G_A(0) = -1.23$, we find, corresponding to the values $M_A = 6, 7, 8, 9m_\pi$, the results for the slope of the photoproduction differential cross section at threshold shown in Table 1, where the experimental results for $(|\kappa|/|q|) \, d\sigma_+/d\Omega$ [20] and σ_-/σ_+ [21] are also quoted. Concerning the latter, however, we have new indications [22] which seem to point towards the value $R = 1.35$.

In the electroproduction case, a recent threshold measurement [29] corresponding to the values

$$\kappa^2 = -10.4 \, m_\pi^2; \quad E_l^L = 800 \text{ MeV};$$
$$E_l'^L = 542 \text{ MeV}; \quad \varepsilon = 0.74 \tag{4.15}$$

Table 1. Slope of the differential cross section at threshold (μb/sr) for π^+-photoproduction and π^+/π^--ratio, assuming $G_A(0) = -1.23$

M_A	$6m_\pi$	$7m_\pi$	$8m_\pi$	$9m_\pi$	exp.
$\dfrac{\|k\|}{\|q\|} \dfrac{d\sigma_+}{d\Omega}$	15.2	15.6	15.8	16.0	15.6 ± 0.5
$R = \sigma_-/\sigma_+$	1.35	1.35	1.34	1.34	1.265 ± 0.075
					1.35

Table 2. π^+-electroproduction at threshold in units of $10^{-31}\,\text{cm}^2\,(\text{GeV}/c)^{-2}\,\text{sr}^{-1}$ and σ_L/σ_T-ratio corresponding to the value $G_A(0) = -1.23$. $\kappa^2 = -10.4\,\mu^2$, $E_l^L = 800\,\text{MeV}$, $E_l'^L = 542\,\text{MeV}$, $\varepsilon = 0.74$.

M_A	$6m_\pi$	$7m_\pi$	$8m_\pi$	$9m_\pi$	exp. [29]
$\dfrac{1}{|q|}\dfrac{d^2\sigma}{d\Omega_l'^L dE_l'^L}$	4.03	5.15	6.00	6.72	4.9 ± 0.7
σ_L/σ_T	0.53	0.40	0.32	0.28	

Table 3. $d\sigma(90°)/d\Omega$ (unpol.), $d\sigma_\perp(90°)/d\Omega$, both in µb/sr, asymmetry ratio Σ and $\sigma_\parallel/\sigma_\perp$ at 90° as functions of photon lab. energy. $G_A(0) = -1.23$, $M_A = 7m_\pi$.

$$d\sigma_\perp/d\Omega = (\alpha M^2 |q|/4\pi W^2 |\kappa|)(A_\perp + B_\perp \cos\theta + C_\perp \cos^2\theta) + O(|q|^3)$$

$$A_\perp = F_1^2 + y^2, \quad B_\perp = B = 2x\bar{E}_{0+}, \quad C_\perp = x^2 - y^2 + 15\bar{E}_{0+}\psi$$

E_γ^L (MeV)	$d\sigma/d\Omega$ unpol. π^+	$d\sigma_\perp/d\Omega$	Σ	$d\sigma_\parallel/d\sigma_\perp$	$d\sigma/d\Omega$ unpol. π^-
160	4.96	5.26	0.059	0.89	6.11
170	6.38	7.09	0.119	0.79	7.80
180	7.17	8.34	0.163	0.70	8.70
190	7.85	9.89	0.192	0.70	9.45
200	8.68	10.33	0.190	0.68	10.36
210	9.47	11.11	0.172	0.70	11.24
220	10.41	11.70	0.134	0.76	12.30

gives the result:

$$\lim_{|q_\pi| \to 0} |q_\pi|^{-1} d^2\sigma(\pi^+)/d\Omega_l'^L dE_l'^L = (4.9 \pm 14\%)\,10^{-31}\,\text{cm}^2(\text{GeV}/c)^{-2}\,\text{sr}^{-1} \tag{4.16}$$

In the same kinematical configuration, our results are summarized in Table 2, where also the σ_L/σ_T ratio is given: From Tables 1, 2 we conclude that the agreement with the available threshold data is excellent if we choose $M_A = 7m_\pi$ (and $G_A(0) = -1.23$). Therefore we fix from now on $M_A = 7m_\pi$ and try a comparison with data beyond threshold. In the photoproduction case we find, at $\theta = 90°$, the results quoted in Table 3 and one can see in Fig. 1 the way our points fit the experimental [22] curve (something like ± 5–10% has to be attached to every point) for π^+. In Fig. 2 the asymmetry ratio [30] is finally given.

In electroproduction, no data are available till now. We give in any case in Table 4 our results for the ratio σ_L/σ_T at threshold at various κ^2, for different values of M_A.

A general (and obvious) comment has to be added to our numerical predictions. They arise, as we already said, from an expression containing

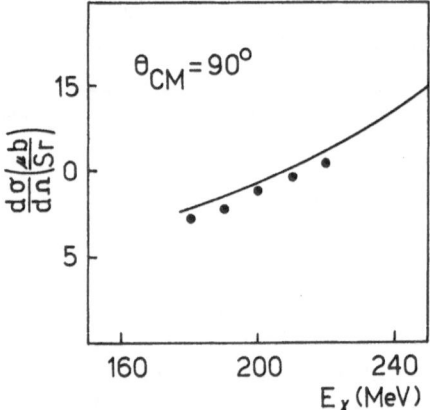

Fig. 1. π^+ photoproduction differential cross section at 90°, as a function of the Lab. photon energy. The dots (●) represent our numerical predictions

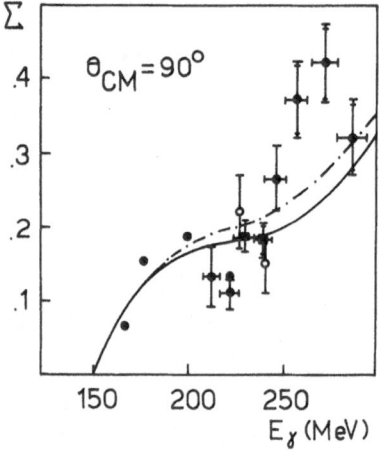

Fig. 2. Asymmetry ratio Σ for π^+ photoproduction at 90°, as a function of the Lab. photon energy. The curves represent dispersive calculations, by Schmidt (———) and by Donnachie and Shaw (—·—·—). The dots (●) represent the theoretical predictions. For the experimenatl points see Ref. [30]

a leading term, with a simple dependence on $G_A(t)$ and $D(t)$, plus a "continuum" part. Once a theoretical estimate of this last contribution is available, one can test from a comparison with experiment the validity of fits like Eqs. (4.13), (4.14) for the axial vector form factors, in particular for $D(t)$. Conversely the excellent agreement obtained in photoproduction, where these fits should be rather good due to the small t value involved, represents a good check of our theoretical computation of the

Table 4. σ_L/σ_T-ratio at threshold, corresponding to the choice $G_A(0) = -1.23$

κ^2	$M_A=6m_\pi$	$M_A=7m_\pi$	$M_A=8m_\pi$	$M_A=9m_\pi$	$M_A=6m_\pi$	$M_A=7m_\pi$	$M_A=8m_\pi$	$M_A=9m_\pi$
	$\sigma_L/\sigma_T(\pi^+)$				$\sigma_L/\sigma_T(\pi^-)$			
$-\ 3m_\pi^2$	0.33	0.30	0.28	0.27	0.53	0.49	0.46	0.44
$-\ 6m_\pi^2$	0.41	0.34	0.30	0.28	0.78	0.67	0.61	0.56
$-\ 9m_\pi^2$	0.47	0.37	0.31	0.27	1.04	0.84	0.72	0.64
$-12m_\pi^2$	0.59	0.42	0.33	0.28	1.30	0.97	0.79	0.69

continuum. However for larger values of t the accuracy of those simple pole expressions could fail, in particular $D(t)$ could be not well accounted for by (4.14). If one believes in the extended validity of our computations, then measurements of σ_T and σ_L at threshold could provide a useful information about the effective t-dependence of these form factors.

The importance of a careful estimate of the continua can be appreciated by comparing our indication $M_A \sim 7m_\pi$ with a recent fit for $G_A(t)$ by Nambu and Yoshimura, obtained by matching experimental data near threshold with an improved version of the original Nambu-Shrauner formula. They obtain $M_A \sim 9.5\,m_\pi$ and the different determination can be ascribed to the different size of the finite pion mass effects (that in their case never exceed 10%).

We are thus led to a quick comparison between our results and other theoretical calculations of low energy electroproduction.

Dispersion Theory

Clearly the dispersion relation approach, based on the solution of multipole integral equations remains the most complete theory. However, the practical determination of the multipoles and their dependence on the input parameters is not always easy and we believe that in the low energy region (below the first resonance) the present approach can be considered as a reasonable alternative to the dispersive scheme. In the photoproduction case the agreement between the two descriptions appears to be quite encouraging, as shown in Table 5, where our predictions are compared with the dispersive calculation of Donnachie and coworkers [31]. As far as electroproduction is concerned, we want to stress that the dispersive approach and the present one exhibit an explicit dependence on different quantities. For instance in the Born approximation of dispersion relations the pion form factor is usually included; it does not appear here[18], but on the other hand $G_A(t)$ and $G_P(t)$ must be supplied.

---

[18] Although there is the photoelectric pion term, accounted for by the pion pole in $D(t)$.

Table 5. y and z at $\kappa^2 = 0$, and in the evaluation of Berends, Donnachie and Weaver. $y = 2M_{1+} + M_{1-}$; $z = E_{1+} - M_{1+}$; $G_A(0) = -1.23$; $M_A = 7m_\pi$

E_γ^L(MeV)	$y^{(-)}$	$y^{(0)}$	$z^{(-)}$	$z^{(0)}$	$y_{\text{B.D.W.}}^{(-)}$	$y_{\text{B.D.W.}}^{(0)}$	$z_{\text{B.D.W.}}^{(-)}$	$z_{\text{B.D.W.}}^{(0)}$
160	−0.136	≃0	0.169	−0.007	−0.150	≃0	0.165	−0.007
180	−0.259	≃0	0.321	−0.013	−0.306	≃0	0.299	−0.013
200	−0.379	≃0	0.470	−0.019	−0.449	≃0	0.405	−0.017
220	−0.489	≃0	0.608	−0.025	−0.590	−0.004	0.496	−0.019

Current Algebra

There are a few recent papers on the subject [32], where a more complete bibliography can also be found. The common idea, whose root is already in the Nambu's papers, is to separate the amplitude into the contribution coming from the Born terms (pion photoelectric diagram included) and into the continuum: $T = T_{\text{B.t.}} + T_{\text{cont}}$. The evaluation of $T_{\text{B.t.}}$ does not offer any difficulty, of course, and it can be done with $m_\pi \neq 0$. In particular as $m_\pi \to 0$, $T_{\text{B.t.}}$ reproduces, for photoproduction, the Kroll-Ruderman theorem. For T_{cont} one can try to use the current algebra constraints, for instance in the form of Eq. (3.4) of p. 130 but these are not enough to determine completely all invariant amplitudes [19]. We can say in general that no explicit estimate of the corrections alternative to our approach has been, at our knowledge, performed.

Appendix

The Cluster Decomposition

The cluster decomposition provides the systematic procedure to identify the various contributions to our completeness sums. The starting point is that every matrix element can be separated into a connected and a disconnected contribution. Indeed, if we consider an amplitude of the form

$$M = \langle p|A|p_1, p_2 \cdots p_n, \beta \rangle \qquad (A.1)$$

where A is a certain operator, $p, p_1 \ldots p_n$ are momenta of identical particles and β represents all other particles appearing in the final state, we have

$$M = M^{\text{conn}} + M^{\text{disc}} \qquad (A.2)$$

where

$$M^{\text{disc}} = \langle p|p_1 \rangle \langle 0|A|p_2 \cdots \beta \rangle$$
$$\pm \langle p|p_2 \rangle \langle 0|A|p_1 \cdots \beta \rangle \pm \cdots \qquad (A.3)$$

and the $\pm$ signs reflect the statistics obeyed by the particle labelled with p.

---

[19] An exception seems to be the work by Scadron and Silbar, where in photoproduction not only $M_1^{(+,0)}$ but also $M_1^{(-)}$, $M_{3,4}$ are claimed to be determined by current algebra.

An explicit form for M^{conn} and M^{disc} can be obtained using the asymptotic creation and destruction operators. The relevant formula is

$$\langle p|A|n\rangle = \langle 0|[a_{\text{out}}(p), A]|n\rangle$$
$$+ \langle 0|A\, a_{\text{out}}(p)|n\rangle\,. \tag{A.4}$$

The first term in the r.h.s. corresponds to the fully connected contribution, as is easily seen using the LSZ reduction technique. The second term represents the disconnected part, since the operator $a(p)$ acts as a destruction operator on the state $|n\rangle = |p_1 \ldots p_n, \beta\rangle$, leading automatically to Eq. (A.3)[20]. In our applications to current algebra we consider only single particle external states:

$$\langle p_2|A|p_1\rangle = \langle p_2|A|p_1\rangle^{\text{conn}} + \langle p_2|p_1\rangle \langle 0|A|0\rangle$$
$$\langle p_2|A|p_1\rangle^{\text{conn}} = \langle 0|[[a(p_2), A], a^+(p_1)]|0\rangle \tag{A.5}$$

and, moreover, $A \equiv A_2 A_1$.

Separating out the fully disconnected part, and expanding the double commutator, we get:

$$M_{21} \equiv \langle p_2|A_2 A_1|p_1\rangle^{\text{conn}}$$
$$= \langle 0|[a_2, A_2][A_1, a_1^+]|0\rangle + \langle 0|A_2[a_2, [A_1, a_1^+]]|0\rangle \tag{A.6}$$
$$+ \langle 0|[[a_2, A_2], a_1^+]A_1|0\rangle \pm \langle 0|[A_2, a_1^+][a_2, A_1]|0\rangle\,,$$

where the $\pm$ signs apply respectively to external bosons and fermions.

At this point we use a completeness, and find:

$$M_{21} = M_{21}^{\text{I}} + M_{21}^{\text{II}} + M_{21}^{\text{III}}\,, \tag{A.7}$$

$$M_{21}^{\text{I}} = \sum_n \langle 0|[a_2, A_2]|n\rangle \langle n|[A_1, a_1^+]|0\rangle$$
$$= \sum_n \langle p_2|A_2|n\rangle^{\text{conn}} \langle n|A_1|p_1\rangle^{\text{conn}} \tag{A.8}$$

$$M_{21}^{\text{II}} = \sum_m \langle 0|[[a_2, A_2], a_1^+]|m\rangle \langle m|A_1|0\rangle$$
$$+ \sum_{m'} \langle 0|A_2|m'\rangle \langle m'|[a_2, [A_1, a_1^+]]|0\rangle$$
$$= \sum_m \langle p_2|A_2|m, p_1\rangle^{\text{conn}} \langle m|A_1|0\rangle \tag{A.9}$$
$$+ \sum_{m'} \langle 0|A_2|m'\rangle \langle m', p_2|A_1|p_1\rangle^{\text{conn}}\,,$$

$$M_{21}^{\text{III}} = \pm \sum_l \langle 0|[A_2, a_1^+]|l\rangle \langle l|[a_2, A_1]|0\rangle$$
$$= \sum_l \langle 0|A_2|l, p_1\rangle \langle l, p_2|A_1|0\rangle\,. \tag{A.10}$$

---

[20] Since $|p\rangle$ is a one particle state, it is inessential to specify whether it is ingoing or outgoing.

In order to understand the meaning of the different terms, let us suppose explicitly $|p_1\rangle$, $|p_2\rangle$ to be nucleon states, and $A_{1,2}$ pseudoscalar operators with zero baryonic number (for example, axial charges). In this case we see that:

1. The states $|n\rangle$ have baryonic number $+1$ (the lowest one is the nucleon state), and M^I_{21} corresponds to direct graphs, namely to the transition from a nucleon state to an intermediate one and again to the final nucleon state.

2. The states $|m\rangle$, $|m'\rangle$ have zero baryonic number, and M^{II}_{21} represents the "semidisconnected graphs" (or "mass singularities"), where, e.g. a state $|m\rangle$ is created from the vacuum and the reaction $m + p_1 \rightarrow A_2 + p_2$ follows. The lowest contribution to Eq. (A.9) will come from the single pion state. Under certain conditions, this will be the dominant contribution to the completeness sum. The results of soft pion theory can thus be derived from axial charge-commutators, under the assumption of pion dominance.

3. The states $|l\rangle$ have baryon number -1 and M^{III}_{21} is given by the contributions of "Z-graphs", corresponding to the creation from the vacuum of the state $|l, p_2\rangle$, with subsequent annihilation of the state $|l, p_1\rangle$ (the simplest effect is the creation and destruction of a nucleon-antinucleon pair).

References

1. de Alfaro, V., Fubini, S., Furlan, G., Rossetti, C.: Nuovo Cimento **62 A**, 497 (1969).
2. Fubini, S., Furlan, G.: Ann. Phys. **48**, 322 (1968).
3. Low, F. E.: Phys. Rev. **97**, 1392 (1955).
4. Li, L. F., Pagels, H.: Phys. Rev. Letters **36**, 1204 (1971).
5. Weinberg, S.: Phys. Rev. Letters **17**, 616 (1966);
   Tomozawa, Y.: Nuovo Cimento **46 A**, 707 (1966).
6. von Hippel, F., Kim, J. K.: Phys. Rev. D **1**, 151 (1970).
7. Hakim, S. J.: University of Michigan, preprint (1971);
   Gensini, P.: University of Lecce, preprint (1971).
8. Ericson, M., Rho, M.: CERN preprint, Ref.Th. 1350 (1971). In this paper the magnitude of the σ-term is derived from low-energy pion-nucleus scattering, making use of the technique described in Part a). The numerical result is in agreement with Kim and von Hippel.
9. Gell-Mann, M., Oakes, R. J., Renner, B.: Phys. Rev. **175**, 2195 (1968);
   Glashow, S., Weinberg, S.: Phys. Rev. Letters **20**, 224 (1968).
10. Cheng, T. P., Dashen, R.: Phys. Rev. Letters **26**, 594 (1971). See also: Altarelli, G., Cabibbo, N., Maiani, L.: Preprint ISS 71/20, Rome (1971).
11. Höhler, G., Jacob, H. P., Strauss, R.: Phys. Letters **35** B, 445 (1971).
12. Altarelli, G., Cabibbo, N., Maiani, L.: Phys. Letters **35** B, 415 (1971);
    Mathur, V. S.: Phys. Rev. Letters **27**, 452 (1971);
    Crewther, R.: Phys. Rev. D **3**, 3152 (1971).

13. See, for instance, Brandt, R. A., Preparata, G.: Phys. Rev. Letters **26**, 1605 (1971), where previous work is quoted. However, the main interest of the authors is in $SU(3) \times SU(3)$ breaking, rather than in pion processes.
14. See, anyway, the references quoted by Adler, S. L.: Ann. of Phys. (N.Y.) **50**, 189 (1968) or by von Gehlen, G.: "Threshold pion electro- and photoproduction", Bonn University PI-80 (1970).
15. Low, F.: Phys. Rev. **110**, 964 (1958).
16. Kroll, N. M., Ruderman, M. A.: Phys. Rev. **93**, 233 (1954).
17. Nambu, Y., Lurié, D.: Phys. Rev. **125**, 1429 (1962);
    Nambu, Y., Shrauner, E.: Phys. Rev. **128**, 862 (1962);
    Nambu, Y., Yoshimura, M.: Phys. Rev. Letters **24**, 25 (1970).
18. Verzegnassi, C.: DNPL/R8.
19. See, for instance, Zagury, N.: Phys. Rev. **143**, 112 (1966).
20. Burq, J. P.: Ann. Phys. (Paris) **10**, 363 (1965).
21. Govorkov, B. B., Denisov, S. D., Minarik, E. V.: J. Nucl. Phys. (U.S.S.R.) **4**, 371 (1966).
22. Noelle, P.: Bonn University, preprint PI 2–92 (1971).
23. Govorkov, B. B., Denisov, S. P., Minarik, E. V.: See Ref. [21] and Sov. J. Nucl. Phys. **4**, 265 (1966).
24. Fubini, S., Furlan, G., Rosetti, C.: Nuovo Cimento **40** A, 1171 (1965).
25. Furlan, G., Jengo, R., Remiddi, E.: Nuovo Cimento **44** A, 427 (1966). See also: Adler, S. L., Gilman, F. J.: Phys. Rev. **152**, 1460 (1966); Riazuddin, Lee, B. W.: Phys. Rev. **146** B, 1202 (1966).
26. Furlan, G., Paver, N., Verzegnassi, C.: Nuovo Cimento X, **63** A, 519 (1969).
27. Gross, D. J., Jackiw, R.: Phys. Rev. **163**, 1688 (1967).
28. Paver, N., Verzegnassi, C.: Nota interna N. AE 69/5 (1970);
    Furlan, G., Paver, N., Verzegnassi, C.: Nuovo Cimento X, **70** A, 247 (1970).
29. Amaldi, E., Borgia, B., Pistilli, P., Balla, M., Di Giorgio, G. V., Giazotto, A., Serbassi, S., Stoppini, G.: Nuovo Cimento **65** A, 377 (1970).
30. Grilli, M., Nigro, M., Schiavon, E., Soso, F., Spillantini, P., Valente, V.: Nota interna N. 356, LNF-67/18 (1967).
31. Berends, F. A., Donnachie, A., Weaver, D. L.: Nucl. Phys. **4** B, 1 (1968).
32. De Baenst, P.: Nucl. Phys. B **24**, 633 (1970);
    Vainshtein, A. I., Zakharov, V. I.: Novosibirsk preprint (1970);
    Scadron, M. D., Silbar, R. R.: University of Arizona preprint (1970). Earlier references are found in these papers.

Professor Dr. G. Furlan, Dr. N. Paver, Dr. C. Verzegnassi
Istituto di Fisica Teorica, University of Trieste, Italy
and Istituto Nazionale di Fisica Nucleare, Sottosezione of Trieste, Italy

Springer Tracts in Modern Physics

To appear in forthcoming volumes:

H. Arenhövel and H. J. Weber: Nuclear Isobar Configurations

J. Brandmüller and R. Claus: Light Scattering on Optical Phonons and Polaritons

R. Graham: Statistical Theory of Instabilities in Stationary Non-Equilibrium Systems with Applications to Lasers and Non-Linear Optics

K. Heinloth: Experiments on Electroproduction in High Energy Physics

D. Schmid: Nuclear Magnetic Double Resonance – Principles and Applications in Solid State Physics

H. Theissen: Spectroscopy of Light Nuclei by Low Energy (70 MeV) Inelastic Electron Scattering

Volume 63

Photon-Hadron Interactions II. Lectures presented at the International Summer Institute in Theoretical Physics. DESY, Hamburg, July 12—24, 1971

A. P. Contogouris: Regge Analysis and Dual Absorptive Model

A. Donnachie: Exotic Electromagnetic Currents

J. Frøyland: High Energy Photoproduction of Pseudoscalar Mesons

F. M. Renard: p-ω Mixing

D. Schildknecht: Vector Meson Dominance, Photo- and Electro-production from Nucleons

K. Schilling: Some Aspects of Vector Meson Photoproduction on Protons

P. D. B. Collins and F. D. Gault: The Eikonal Model for Regge Cuts in Pion-Nucleon Scattering

Volume 64

T. Springer: Quasi-Elastic Neutron Scattering for the Investigation of Diffuse Motions in Solids and Liquids

Distributor in USA:

Springer-Verlag New York, Inc.
175 Fifth Ave, New York, N. Y. 10010

SPRINGER TRACTS IN MODERN PHYSICS

Ergebnisse
der exakten Natur-
wissenschaften

Volume **62**

Reprint

R. Jackiw

**Canonical Light-Cone Commutators
and Their Applications**

Springer-Verlag Berlin Heidelberg New York 1972

SPRINGER-VERLAG
BERLIN·HEIDELBERG·NEW YORK

Springer Tracts in Modern Physics

SPRINGER TRACTS
IN MODERN PHYSICS

Ergebnisse
der exakten Natur-
wissenschaften

Volume **62**

Reprint

H. D. Dahmen

Local Saturation of Commutator Matrix Elements

Springer-Verlag Berlin Heidelberg New York 1972

SPRINGER TRACTS
IN MODERN PHYSICS

Ergebnisse
der exakten Natur-
wissenschaften

Volume 62

Reprint

P. V. Landshoff
Duality in Deep Inelastic Electroproduction

Springer-Verlag Berlin Heidelberg New York 1972

SPRINGER TRACTS IN MODERN PHYSICS

Ergebnisse
der exakten Natur-
wissenschaften

Volume **62**

Reprint

C. H. Llewellyn Smith
Parton Models of Inelastic Lepton Scattering

Springer-Verlag Berlin Heidelberg New York 1972

SPRINGER TRACTS IN MODERN PHYSICS

Ergebnisse
der exakten Natur-
wissenschaften

Volume **62**

Reprint

H. R. Rubinstein
Duality for Real and Virtual Photons

Springer-Verlag Berlin Heidelberg New York 1972

SPRINGER TRACTS
IN MODERN PHYSICS

Ergebnisse
der exakten Natur-
wissenschaften

Volume **62**

Reprint

V. Rittenberg
Scaling in Deep Inelastic Scattering
with Fixed Final States

Springer-Verlag Berlin Heidelberg New York 1972

SPRINGER TRACTS IN MODERN PHYSICS

Ergebnisse
der exakten Natur-
wissenschaften

Volume **62**

Reprint

K. Huang
Duality and the Pion Electromagnetic Form Factor

Springer-Verlag Berlin Heidelberg New York 1972

SPRINGER TRACTS IN MODERN PHYSICS

Ergebnisse
der exakten Natur-
wissenschaften

Volume **62**

Reprint

K. Huang
**Deep Inelastic Hadronic Scattering
in Dual-Resonance Model**

Springer-Verlag Berlin Heidelberg New York 1972

SPRINGER TRACTS
IN MODERN PHYSICS

Ergebnisse
der exakten Natur-
wissenschaften

Volume **62**

Reprint

G. Furlan, N. Paver and C. Verzegnassi

**Low Energy Theorems and Photo- and Electro-
production Near Threshold by Current Algebra**

Springer-Verlag Berlin Heidelberg New York 1972